JOURNAL OF GREEN ENGINEERING

Chairperson: Ramjee Prasad, CTIF, Aalborg University, Denmark
Editor-in-Chief: Dina Simunic, University of Zagreb, Croatia

Editorial Board
Luis Kun, Homeland Security, National Defense University, i-College, USA
Dragan Boscovic, Motorola, USA
Panagiotis Demstichas, University of Piraeus, Greece
Afonso Ferreira, CNRS, France
Meir Goldman, Pi-Sheva Technology & Machines Ltd., Israel
Laurent Herault, CEA-LETI, MINATEC, France
Milan Dado, University of Zilina, Slovak Republic
Demetres Kouvatsos, University of Bradford, United Kingdom
Soulla Louca, University of Nicosia, Cyprus
Shingo Ohmori, CTIF-Japan, Japan
Doina Banciu, National Institute for Research and Development in Informatics, Romania
Hrvoje Domitrovic, University of Zagreb, Croatia
Reinhard Pfliegl, Austria Tech-Federal Agency for Technological Measures Ltd., Austria
Fernando Jose da Silva Velez, Universidade da Beira Interior, Portugal
Michel Israel, Medical University, Bulgaria
Sandro Rambaldi, Universita di Bologna, Italy
Debasis Bandyopadhyay, TCS, India

Aims and Scopes
Journal of Green Engineering will publish original, high quality, peer-reviewed research papers and review articles dealing with environmentally safe engineering including their systems. Paper submission is solicited on:

- Theoretical and numerical modeling of environmentally safe electrical engineering devices and systems.
- Simulation of performance of innovative energy supply systems including renewable energy systems, as well as energy harvesting systems.
- Modeling and optimization of human environmentally conscientiousness environment (especially related to electromagnetics and acoustics).
- Modeling and optimization of applications of engineering sciences and technology to medicine and biology.
- Advances in modeling including optimization, product modeling, fault detection and diagnostics, inverse models.
- Advances in software and systems interoperability, validation and calibration techniques. Simulation tools for sustainable environment (especially electromagnetic, and acoustic).
- Experiences on teaching environmentally safe engineering (including applications of engineering sciences and technology to medicine and biology).

All these topics may be addressed from a global scale to a microscopic scale, and for different phases during the life cycle.

JOURNAL OF GREEN ENGINEERING

Volume 1   No. 3   April 2011

Editorial Foreword	v–vi
P. LINDGREN and Y. TARAN / A Futuristic Outlook on Business Models and Business Model Innovation in a Future Green Society	229–239
L. GAVRILOVSKA, V. RAKOVIC and V. ATANASOVSKI / Energy Efficiency of Resource Management Architectures in Heterogeneous Wireless Networks	241–254
P. DONADIO, A. CIMMINO, and R. PRASAD / A Cloud Infrastructure to Manage Future Internet: The Virtual Network Operation Center	255–265
EMILIO CALVANESE STRINATI, ANTONIO DE DOMENICO, and LAURENT HERAULT / Green Communications: An Emerging Challenge for Mobile Broadband Communication Networks	267–301
POUL EJNAR ROVSING, PETER GORM LARSEN, THOMAS SKJØDEBERG TOFTEGAARD and DANIEL LUX / A Reality Check on Home Automation Technologies	303–327
NUNO COUTINHO, TIAGO CONDEIXA, SUSANA SARGENTO and AUGUSTO NETO / Energy Efficiency as Input for Context-Aware Group-Based Communications	329–353

Published, sold and distributed by:
River Publishers
P.O. Box 1657
Algade 42
9000 Aalborg
Denmark

Tel.: +45369953197
www.riverpublishers.com

Journal of Green Engineering is published four times a year.
Publication programme, 2010–2011: Volume 1 (4 issues)

ISSN 1904-4720

All rights reserved © 2011 River Publishers

No part of this work may be reproduced, stored in a retrieval system, or transmitted in any form or by any means, electronic, mechanical, photocopying, microfilming, recording or otherwise, without prior written permission from the Publisher.

Editorial Foreword

Dina Simunic[1] and Ramjee Prasad[2]

[1]*University of Zagreb, Croatia*
[2]*Aalborg University, CTIF, Denmark*

Dear Reader,

We are very happy to announce the third issue of the Journal of Green Engineering. This issue presents six papers, which all cover very interesting and timely topics in the field of green engineering. The first paper by P. Lindgren and Y. Taran, entitled "A Futuristic Outlook on Business Models and Business Model Innovation in a Future Green Society" portrays possibilities in the development of green oriented business models, and clarifies the processes of business model innovation for the green society of the future. The second paper, by L. Gavrilovska, V. Rakovic, and V. Atanasovski, entitled "Energy Efficiency of Resource Management Architectures in Heterogeneous Wireless Networks" gives an insight into the energy consumption issue in heterogeneous wireless systems; their analysis focuses on the IEEE 802.21 standard and two different types of RM architectures throughout extensive simulations. The paper by P. Donadio, A. Cimmino, and R. Prasad: "A Cloud Infrastructure to Manage Future Internet: The Virtual Network Operation Center" shows how the setting up and operation of the Virtual NOC should be enabled by emerging technologies for mobile and fixed network business and related technologies. Emilio Calvanese Strinati, Antonio De Domenico, and Laurent Herault, in their paper "Green Communications: An Emerging Challenge for Mobile Broadband Communication Networks" reveal how proposed methodologies permit to achieve notable energy gain over traditional resource allocation techniques, especially in not saturated scenarios, whereas Poul Ejnar Rovsing, Peter Gorm Larsen, Thomas Skjødeberg Toftegaard, and Daniel Lux, in "A Reality Check on Home Automation Technologies" report some of the remaining challenges and the future trends in home automation technologies. The final paper in the issue, by Nuno Coutinho, Tiago Condeixa, Susana Sargento, and Augusto Neto on "Energy Efficiency as Input for Context-Aware Group-Based Communications" shows that context-based grouping and network selection are able to improve users' quality of experience.

In addition to presenting this issue, we are pleased to announce the *34th International Convention on Information and Communication Technology, Electronics and Microelectronics – MIPRO 2011*, to be held in Opatija/Abbazia, Croatia, on the Adriatic Coast from 23–27 May 2011, where the main topic will be "The Green ICT World". With respect to the content and number of participants MIPRO is the oldest and the most wide-ranging ICT convention in Southeast Europe and the Mediterranean. It is the traditional meeting point for experts in the field of economy, education, science, state administration and local government. Because of its synergic activity, MIPRO is not only valuable to the industry but to society in general as well. MIPRO conventions are attended by experts from more than 30 countries with the purpose to exchange knowledge and experience in their field of expertise. Their achievements are presented by means of scientific and professional papers, demonstrations of particular technologies and technical solutions or discussions at round tables and workshops. The participation of major companies to this convention is particulary important because MIPRO allows them to present their state-of-the-art technologies. More information can be found on www.mipro.hr.

The core of "The Green ICT World" consists of three special workshops, entitled "Eco-ICT for Sustainability Development", "Eco-Efficient ICT" and "Smart Cars, Smart Roads". Thus, the *Journal of Green Engineering* is the official journal and media sponsor for MIPRO. Approximately 1000 participants are expected.

We very much hope to see you in Opatija during the 34th MIPRO and thank you all very much for your kind support of the green idea and for your active cooperation!

Respectfully,

Dina and Ramjee

A Futuristic Outlook on Business Models and Business Model Innovation in a Future Green Society

P. Lindgren and Y. Taran

Centre for Industrial Production, Aalborg University, 9220 Aalborg, Denmark;
e-mail: {pel, yariv}@production.aau.dk

Received: 17 December 2010; Accepted: 15 February 2011

Abstract

Companies nowadays invest more capital and resources in green technologies and increasingly start to think more radically when considering their business model innovation processes. However, the development and innovation of green business models to the so called green society is a complex venture and has not been widely researched yet.

The study of green oriented innovation has intensified tremendously in the last decade due to both user demand ('pull') and the development of new mobile and wireless communication technologies ('push'), which gives hope in bringing the green society vision into reality.

Accordingly, this paper portrays a conceptual futuristic outlook into the development of green oriented business models, and clarifies the processes of business model innovation for the future green society.

Keywords: business model, innovation, green business models, green society.

1 Introduction

In 2010 we are in the midst of 3G. Thanks to heavy research and big investments in mobile and wireless technology we are now heading quickly towards 4G and 5G. WWWW: World Wide Wireless Web will emerge soon; its evolution will be based on 4G and will result in a completely wireless and interconnected world. 5G will have an even more important impact on the industry and will add more services and benefits to the world.

Also, given the fact that the 5G will encompass a more intelligent technology system, it is expected that it will open the possibility for new, perhaps disruptive, green business models and will create more intelligent business models that will enhance the interconnection even more, as well as the interdependency between companies and people around the globe.

Yet, from a Business Model (BM) perspective companies are still facing tremendous challenges in understanding the processes of reshaping their existing BM into a new, green oriented one. Consequently, in many cases, and despite the fact that many managers can observe the enormous potential existed in innovating green BMs today, they still find those processes to be highly risky and also somewhat foreign to the existing innovation processes which they have been experimenting with in the past (e.g. product innovation process). Consequently, their existing business model is hardly being reviewed or changed dramatically into the direction of what we call the "green society".

2 Components of the Business Model

The term BM became popular in the mid-1990s during the "dot com era". As business ecosystems emerged, many companies started to rethink their business model and business structure by shifting to an ICT based business model or an "E-form" business [19]. Many authors (see e.g. [29]) have attempted to define the business model concept. It seems that most (if not all) authors agree that a business model is simply a combination of two terms: "business" and "model".

Accordingly, a company's "business model" serves as a building platform that represents the company's operational and physical manifestation. Thus, the challenge for business model "designers" is to first identify the key elements and the key relationships that describe the company's "AS-IS" business model before innovating it.

Building on various studies that have been carried out between the late 1990s and 2003, Morris [20] tried to build what is called "a unified perspect-

Table 1 Core components of the business model.

Core question	Core building block
Who do we serve?	*Target customer*/s, market segments and geographies.
What do we provide?	*Value proposition*/s (products, services and processes) that the company offers.
How do we provide it?	*Value chain configuration* (internal)
	Core competences (assets, processes and activities) that translate companies' inputs into value for customers (outputs).
	Partner network: both strategic partnerships and supply chain management.
	Relationship/s (e.g. physical, digital, virtual, personal, peers, mass awareness).
How do we make money?	*Profit formula* – both turnover and cost structure together with revenue flow.

ive of business models". The authors argued that a business model framework must be reasonably simple, logical, measurable, comprehensive, operational and meaningful. Although Morris et al., as well as other research groups (see e.g. [22]), have systematically analyzed relatively the same list of authors [2–5, 9, 11, 15, 17, 18, 23, 27, 28, 30] they reached different conclusions. Despite the large variation in opinions we could still identify a strong resemblance between the different components. Thus, based mostly on Osterwalder et al.'s [22] nine building blocks, Amit and Zott's [3] analysis, Chesbrough's [6] open business model innovation, Johnson et al. [14], and Hamel [11], we propose the following seven building blocks (Table 1) to best represent the core components of a business model.

3 When Is a BM New?

The next issue concerns the question when we can call a change in the model a business model innovation. Three approaches have been proposed. The first approach "defines" business model innovation as a radical change in the way a company does business [6, 12, 13, 15]. Linder and Cantrell, in particular, clearly attempt to draw a line between what can be defined as business model innovation and what cannot. The second approach regards any change in any of the [core] building blocks or the relationships between them as a form of business model innovation [3, 17, 22]. The third approach, in line with Abell [1] and Skarzynski and Gibson [26], involves considering the number of building blocks that are changed. Any change in one of the building blocks

Table 2 Incremental and radical orientation to each building block.

Building block	Incremental innovation "Do what we do but better"	Radical innovation "Do something different"
1. Value proposition	Offering "more of the same"	Offering something different (at least to the company)
2. Target customer	Existing market and customers	New market and new customers
3. Value chain architecture	Exploitation (e.g. internal, lean, continuous improvements)	Exploration (e.g. open, flexible, diversified)
4. Competences	Familiar competences (e.g. improvement of existing technology, HR, organizational systems, culture)	Disruptively new, unfamiliar, competences (e.g. new, different, emerging technology, HR, organizational systems, culture)
5. Partner network	Familiar (fixed) network	New (dynamic) networks (e.g. alliance, joint-venture)
6. Relationships	Continuous improvements of existed relationships	New relationships (e.g. physical, virtual, personal)
7. Profit formula	Existing processes to generate revenues followed-by/or incremental processes of retrenchments and cost cutting	New processes to generate revenues followed-by/or disruptive processes of retrenchments and cost cutting

would then constitute an incremental innovation. Changes in all the building blocks would be the most radical form of business model innovation.

Another approach defines innovativeness in terms of, what might be called, the reach of the innovation (e.g. [8, 10, 21, 25]). A suitable scale to measure the "new to whom" of a company's innovations could be one ranging from new to the company, via new to the market and new to the industry, to new to the world. If we combine all approaches, a three-dimensional space emerges (Figure 1), which helps in qualifying the innovativeness of a new business model:

- *Radicality* – (how new?) incremental vs. radical of each building block (illustrated in Table 2).
- *Reach* – to whom the innovation is new?
- *Complexity* – number of building blocks changed simultaneously.

Consequently, we get around the eternal discussion of when a BM innovation is indeed radical or incremental, simple or complex, far reaching or not, and, instead, portray the space in which any business model innovation can be positioned in terms of its degree of innovativeness by means of its radically, reach and complexity (Figure 2).

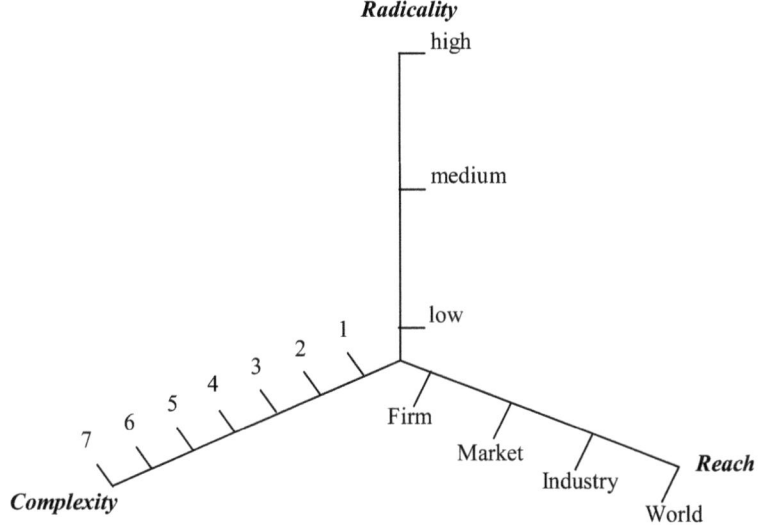

Figure 1 A three-dimensional business model innovation scale [29].

4 Towards an Open BM

Recent research has suggested that, given today's competition, it is highly unlikely that a single company would be able to possess all the necessary competencies needed in order to deliver successfully their unique solutions into the market place, particularly through breakthrough BM innovations (e.g. [6, 16]). Consequently, in order for companies to secure continuous innovation capabilities, it is recommended for companies to involve a bigger group of stakeholders through their innovation processes.

Chesbrough [5, 6], in particular, argued that businesses must adopt a model of innovation that looks both outside of its own four walls for ideas, as well as licenses its home-grown but unused intellectual property to others. That is, in order to truly exploit the true potential of the innovation capabilities of the firm, companies must open their business models by actively searching for and exploiting outside ideas and by equally allowing unused internal technologies to flow to the outside, where other firms can unlock their latent economic potential. For instance, one company develops a novel idea but does not bring it to market. Instead, the company decides to partner with or sell the idea to another party, which then commercializes it.

234 P. Lindgren and Y. Taran

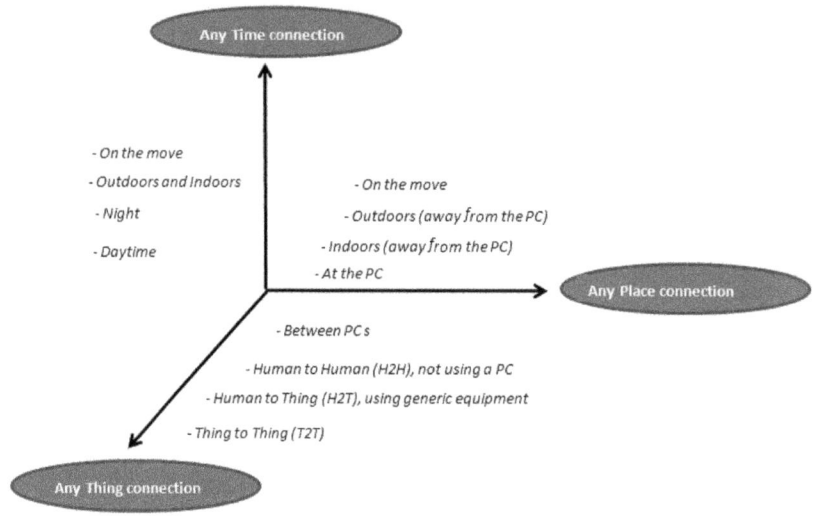

Figure 2 A three-dimensional model for future business model innovation (inspired by a TU Delft presentation ITU adapted from Nomura Research Institute [24]).

5 ICT as a Key Enabler in New BMs in the Green Society

Information and communication technologies appear to be of ever increasing importance to innovation, and will therefore provide the backbone of green business models in the future green society. Green based innovation will increasingly rely on, and will be enabled by, ICT. Innovation will be carried out via advanced ICT tools, and will facilitate the connectivity of:

- Anything – human to human; human to machine; machine back to human; machine to machine.
- Any time – day or night.
- Any place – around the globe.

These new business models will be based on networks, and will encompass different ICT platforms, competences and innovation participants. All of which, will be able to open new possibilities for the development of green-based technologies developments, and BM innovations.

Digital BM types, which include: computers digital networks, network of computers and cloud of computers and people, will all integrate. Virtual business models will include both single company business model, and multi business models working from different locations simultaneously (e.g. [7, 16, 31]). In this context, we believe that the green profile will indeed be

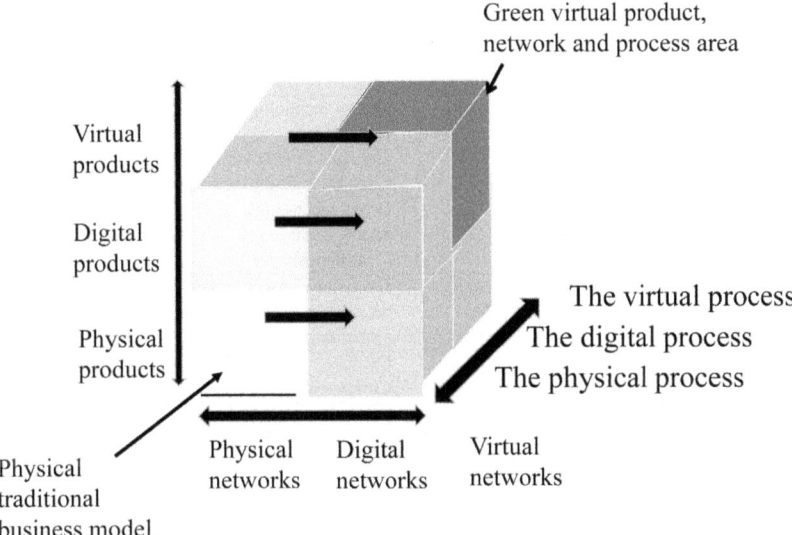

Figure 3 From physical to digital to green virtual business model based on networks (Lindgren [16] inspired by Whinston et al. [31]).

important, by adding more value to the customer and thereby give a better competitive positioning to companies.

The green business models of the future will not have a (purely) physical, digital or virtual character, but rather a combination of these, in a continuous integrated and connected process, wherever and whenever the customer demands it. Also, the integration of all three characteristics, followed by the advance ICT green-based innovations, will provide the ability to manage the mega information flow on the net more adequately, in an agile, flexible and secure way. This will provide a better platform for companies and networks to remain competitive, and to keep the business models lively.

6 Conclusion

The green business model concept continues to evolve and embrace new perceptions, challenges and opportunities. As part of the authors' preliminary research, some important trends and characteristics were found, as shown in Table 3.

Table 3 Future trends to green BMs and BM innovation.

Context for innovation	Past trends	Trends for the Future Green BM
Market	National Stable Common Mainly physical customer relation	Global Fragmented, Dynamic, Customized New markets (Blue Ocean) More digital and virtual customer relations Green Society
Technology	Single technology Expensive Data power low Stable On and off	Mix of technology or multi-technology Cheap Data power over capacity Unstable – Rapid new technology changes Green technology Secure technology Continuously on
Network	Closed networks, local networks Fixed networks	Open networks, dynamic networks, agile network, virtual networks, global networks
Companies' Competences	Stable competences developed inside the company or in a narrow networks	Dynamic – flexible competences Competences continuously developed under pressure Competences developed with many network partners – sharing core competences and skills in the innovation process (to reduce the risk of disruptive technological changes within the industry)
Products	Mostly physical products to some extent immaterial products Stable product – long life cycle Limited distribution and marketing channels	A mixture of physical, immaterial, digital and virtual products, service and process that are green Continuous development of product, service and processes – short life cycles
Business model Innovation process	Stable models Slow, linear innovation process	Many business model innovation models (flexible models, dynamic, agile models, learning by doing, using, interacting). Rapid prototyping and simulation of business models Lean and green business model innovation process
Success criteria	Individual success, innovation speed, time to market, cost and performance, local market Emphasis on short term success criteria More emphasis on continuous improvements and managing tangible assets efficiently.	Network-based success, right-speed innovation, time to market, cost and performance, global markets Emphasis on sustainability – short and long term success criteria. More emphasis on radical innovation More emphasis on managing intangible assets efficiently Scalable Business models Business models being green Society based success

This tendency, if continued, will have tremendous consequences on the way that companies will compete in the future. Consequently, open, multi and green business model innovation initiatives are expected to grow significantly in volume. In addition, the development towards network-based innovation may result in a radical change of our customer focus and understanding, since it (potentially) involves new technologies, new value propositions, new value chains, new network formations, and new markets and new customers. Furthermore, the new network formation ties will not necessarily be characterized by its industrial homogeneity, but rather by the large diversification of its network partners' identity for the purpose of pursuing radical innovation possibilities.

References

[1] D.F. Abell. *Defining the Business: The Starting Point of Strategic Planning*. Prentice-Hall, 1980.
[2] A. Afuah and C. Tucci. *Internet Business Models and Strategies*. McGraw Hill, Boston, 2003.
[3] R. Amit and C. Zott. Value creation in e-business. *Strategic Management Journal*, 22(6/7):493–520, 2001.
[4] L.M. Applegate. E-business models: Making sense of the internet business landscape. In G. Dickson, W. Gary and G. DeSanctis (Eds.), *Information Technology and the Future Enterprise: New Models for Managers*. Prentice Hall, Upper Saddle River, NJ, 2001.
[5] H. Chesbrough and R.S. Rosenbloom. The role of the business model in capturing value from innovation: Evidence from XEROX Corporation's technology spinoff companies. *Industrial and Corporate Change*, 11(3):529–555, 2000.
[6] H. Chesbrough. Open Business Models. *How to Thrive in the New Innovation Landscape*. Harvard Business School Press, 2006.
[7] J. Child and D. Faulkner. *Strategies of Co-operation – Managing Alliances, Networks, and Joint Ventures*. Oxford: Oxford University Press, Oxford, 1998.
[8] R. Garcia and R. Calantone. A critical look at technological innovation typology and innovativeness terminology: A literature review. *Journal of Product Innovation Management*, 19(2):110–132, 2002.
[9] J. Gordijn. Value-based requirements engineering – Exploring innovative E-commerce ideas. PhD Thesis, Vrije Universiteit Amsterdam, the Netherlands, 2002.
[10] S.G. Green, M.B. Gavin, and L. Aiman-Smuth. Assessing a multidimensional measure of radical technological innovation. *IEEE Transactions in Engineering Management*, 42(3):203–214, 1995.
[11] G. Hamel. *Leading the Revolution*. Harvard Business School Press, Boston, 2000.
[12] IBM. Expanding the innovation horizon, http://www-935.ibm.com/services/uk/bcs/html/t_ceo.html, 2006.
[13] IBM. The enterprise of the future, http://www-935.ibm.com/services/us/gbs/bus/html/ceostudy2008.html, 2008.

[14] M.W. Johnson, M.C. Christensen, and H. Kagermann. Reinventing your business model. *Harvard Business Review*, 86(12):50–59, 2008.
[15] J. Linder and S. Cantrell. *Changing Business Models: Surveying the Landscape.* Accenture Institute for Strategic Change, Cambridge, 2000.
[16] P. Lindgren, Y. Taran, and H. Boer. From single firm to network-based business model innovation. *International Journal of Entrepreneurship and Innovation Management*, 12(2):122–137, 2009.
[17] J. Magretta. Why business models matter? *Harvard Business Review*, 80(5):86–92, 2002.
[18] B. Mahadevan. Business models for internet-based e-commerce. An anatomy. *California Management Review*, 42(4):55–69, 2000.
[19] J.F. Moore. The new corporate form. In D. Tapscott, A. Lowy, and D. Ticoll (Eds.), *Blueprint to the Digital Economy – Creating Wealth in the Era of E-Business*. McGraw-Hill, New York, 1998.
[20] M. Morris, M. Schmindehutte, and J. Allen. The entrepreneur's business model: Toward a unified perspective. *Journal of Business Research*, 58(6):726–735, 2003.
[21] E.M. Olsen, O.C. Walker, and R.W. Ruekert. Organizing for effective new product development: The moderating role of product innovativeness. *Journal of Marketing*, 59:48–62, 1995.
[22] A. Osterwalder, Y. Pigneur, and L.C. Tucci. Clarifying business models: Origins, present, and future of the concept. *Communications of AIS*, 16:1–25, 2004.
[23] O. Petrovic, C. Kittl, and R.D. Teksten. Developing business models for e-business. In *Proceedings of the International Conference on Electronic Commerce*, Vienna, October–November 2001.
[24] Presentation given by TU Delft at the Strategic workshop on sensor network at Florence July 2010. Inspired by ITU adapted from Nomura Research Institute.
[25] E.M. Rogers. *Diffusion of Innovations*, 3rd ed. The Free Press, New York, 1983.
[26] P. Skarzynski and R. Gibson. *Innovation to the Core*. Harvard Business School Publishing, Boston, 2008.
[27] P. Stahler. Geschäftsmodelle in der digitalen Ökonomie. Merkmale, Strategien und Auswirkungen. PhD Thesis, University of St. Gallen, Switzerland, 2001.
[28] D. Tapscott, D. Ticoll, and A. Lowy. *Digital Capital. Harnessing the Power of Business Webs*. Harvard Business School Press, Boston, 2000.
[29] Y. Taran, H. Boer, and P. Lindgren. Theory building – Towards an understanding of business model innovation processes. In *Proceedings of the International DRUID-DIME Academy Winter Conference, Economics and Management of Innovation, Technology and Organizational Change*, 2009.
[30] P. Weill and M.R. Vitale. *Place to Space*. Harvard Business School Press, Boston, 2001.
[31] A.B. Whinston, D.O. Stahl, and S. Choi. *The Economics of Electronic Commerce*. Macmillan Technical Publishing, Indianapolis, IN, 1997.

Biographies

Peter Lindgren is Associated Professor of Innovation and New Business Development at the Center for Industrial Production at Aalborg University and Manager for International Center for Innovation. He holds a BSc in Business Administration and an MSc in Foreign Trade and PhDs in both Network Based High Speed Innovation. He has (co-)authored numerous articles and several books on Product Development in Network, Electronic Product Development, New Global Business Development, Innovation Management and Leadership and High Speed Innovation. His current research interest is New Global Business models, i.e. the typology and generic types of business models and how to innovate these.

Yariv Taran is a PhD fellow at the Center for Industrial Production at Aalborg University. His current research interest lies in the area of business model innovation. Other areas of research interests include intellectual capital management, knowledge management, entrepreneurship and regional systems of innovation.

Energy Efficiency of Resource Management Architectures in Heterogeneous Wireless Networks

L. Gavrilovska, V. Rakovic and V. Atanasovski

Faculty of Electrical Engineering and Information Technologies, Ss Cyril and Methodius University Skopje, Rugjer Boskovic bb, 1000 Skopje, Macedonia;
e-mail: {liljana, valentin, vladimir}@feit.ukim.edu.mk

Received: 21 December 2010; Accepted: 25 February 2011

Abstract

The Heterogeneous Wireless Networks (HWN) paradigm is based on the interoperability and coexistence of different types of wireless access networks in a unified wireless heterogeneous platform. A key aspect in this concept is the management of the available resources, in terms of user perception, in the most efficient way. Therefore, Resource Management (RM) architectures are tailored to maximize the number of served users and user perception based on the perceived Quality of Service (QoS). Additionally, the energy efficiency and the efficient energy consumption management are relevant aspects in multi-interface mobile terminals due to the multiple active radio interfaces. This paper gives an insight into the energy consumption issue in heterogeneous wireless systems. The analysis focuses on the IEEE 802.21 standard and two different types of RM architectures throughout extensive simulations. The simulation results give clear representation of the energy consumption in the mobile terminals for the IEEE 802.21 and the two different types of RM architectures. The results can be used as initial guidelines for future development of energy aware resource management schemes.

Keywords: energy efficiency, resource management, heterogeneous wireless networks, IEEE 802.21.

1 Introduction

Future wireless networks target integration of existing wireless technologies into a single heterogeneous platform, providing transparency to end users [1, 2]. The Resource Management (RM) plays a key role in the Heterogeneous Wireless Networks (HWN) platform ensuring service continuity of active sessions and network transparency in terms of user perception and perceived QoS level. In this manner, the lately emerging IEEE 802.21 standard [3] is specifically designed to deal with Vertical Handovers (VHO) in HWNs in a seamless fashion. Also, this standard provides valuable OSI 2.5-layer support for designing intelligent RM mechanisms on top (both on network and terminal side).

Energy efficiency is becoming a hot topic in the area of future wireless networks. A variety of mechanisms and protocols that decrease energy consumption in future wireless systems are developed and tested. On the other hand, future mobile terminals are being designed as multi interface nodes, which results in increased number of physical interfaces yielding increased energy consumption. However, a limiting factor can be the limited battery capacity. Therefore, the RM in future HWNs should be an energy aware entity. It should take the energy consumption in consideration while selecting an access network, hence preserving the needed QoS for the active sessions. For example, a distinct feature of the IEEE 802.21 standard is the network awareness and network selection, while using only one active physical interface [3], leading to a decreased energy consumption in multi-interface nodes [4].

In this paper we investigate the behaviour of the IEEE 802.21 standard and two different RM architectures by analyzing their impact on the energy consumption in the mobile terminals. Analysis is performed throughout extensive simulations in QualNet environment that has the ability to simulate the energy consumption of the mobile nodes. The results give a clear insight on the performance of the IEEE 802.21 protocol and the analyzed RM architectures.

The paper is organized as follows: Section 2 gives a short overview of the work done in the area of energy efficient communications in HWN is given. In Section 3 we explain the features that have the main impact on the energy consumption and defines the RM architectures that are analyzed in the paper. In Section 4 we present performance analysis and comparison of the analyzed RM architectures and highlights the benefits of the implementation of IEEE

802.21 standard in terms of energy consumption. Finally, in Section 5 we draw some important conclusions and milestones for future work.

2 Related Work

Every constituent wireless technology of the heterogeneous wireless system has its own energy consumption characteristic. The energy consumption is studied and analyzed based on analytical modelling and measurements. The analytical model in [5] analyzes and quantifies the energy consumption for streaming and bursty types of traffic for WCDMA. Balasubramanian et al. [6] present a comparative measurement study of the energy consumption for three different technologies (WiFi, GSM and UMTS). Another approach of measuring the consumed energy is by investigating the activity of the different operational modes on a given interface [7]. The results in [7] show that the IEEE 802.11 interface has a complex range of behaviours in terms of consumed energy. Multi-interface nodes consume more energy than the single-interface nodes due to the multiple active wireless interfaces. Lee and Golmie [8] propose a novel scheme that uses WWAN (Wireless Wide Area Networks) interface for the tracking and authentication of the WLAN (Wireless Local Area Connection) interface. The WLAN interface can be switched off when it is not used, resulting in smaller energy consumption. The same idea from [8] has been used as an option in the IEEE 802.21 standard. Desset et al. [9] show that while using the MIIS (Media Independent Information Service) from the IEEE 802.21 standard substantial energy efficiency can be achieved. Due to the high power consumption while executing handovers, Desset et al. [9] show that the initialization of the handovers process should occur in a sparser manner in order to achieve energy efficiency in the mobile terminal. In [10] Minji et al. propose a novel RAT selection algorithm that selects the most energy efficient network taking the energy consumption and the network throughput into consideration.

3 Features and Applications

Future RM will have to address the energy efficiency in order to contribute to the "green communication" [11] paradigm. Investigating the cause of increased consumption in the mobile terminals based on the RM functionalities is a valuable process that can lead to designing energy aware RM architectures. This section explains the impact of some distinct features of the RM and the impact of typical RM architectures on the energy consumption.

3.1 Resource Management Features

The resource management module can be based on multiple features that provide insight on the user and network behavior. Some of the features that have impact on the energy consumption of the mobile user are:

- Mobile terminal speed
- Application bitrate
- Channel state
- Network state
- Number of users
- Technology type

The *mobile terminal speed* directly affects the incidence of the handover initialization process. Increased mobile speed will increase the number of initiated handovers resulting in increased energy consumption. Smart management of the handover initialization process while traveling at high speeds can result in decreased energy consumption [12].

Increasing the *application bitrate* increases the consumed energy in the mobile terminal. This parameter can be valuable to the RM in case when energy efficiency is needed. Decreasing the application rate will decrease the energy consumption thus extend the battery lifetime.

The *channel state* can have a huge impact on the energy consumption in the mobile terminal. Bad channel conditions can result in increased transmit power that leads in increased energy consumption. Often it is much more effective (in terms of the energy) to reconnect to another Point of Attachment that has a better channel state.

When the *network* is in a *state* of congestion then retransmissions start to occur frequently. The retransmissions have negative impact on the energy increasing the consumption as the number of the retransmitted datagrams increases. The congestion effect is tightly correlated with the *number of users* in the network, where larger number of users increases the probability of network congestion.

Every *type* of wireless *technology* has a distinct energy consumption behavior. As shown in [6], WiFi has a lower consumption rate compared to UMTS. Choosing the network based on its technology can lead to substantial energy savings in given situations.

The influence of some of the features, mentioned in this subsection, on the energy efficiency will be analyzed through simulations in Section 4.

3.2 Resource Management Architectures

This subsection focuses on two general types of RM architectures for HWN. Both of the architectures are designed to be implemented on the mobile terminal side. The first one is based on a single parametric decision making process for network selection. The most commonly used parameters in the single parametric RM are based on the channel state [13] or the type of the active application [14]. In this paper, the single parametric RM is an SNR (Signal to Noise Ratio) based RM that selects the most suitable network based on its channel SNR level. Its choice for the comparison analysis is based on its wide usage in today's wireless and mobile networks.

The second type of architectures is based on multiple parametric decision making process for the network selection. They consume more energy in terms of processing power, but provide better performance to the mobile users. This type of RM architectures can be additionally divided based on the type of the decision algorithm, as non cognitive based on predefined policies, i.e. mathematical relations, and cognitive based on Artificial Intelligence (AI) [15]. In this paper, the multi parametric RM is based on complex mathematical functions that are used for selecting the most suitable network [16]. The next section analyzes the performance of the two RM architectures in terms of energy consumption in the mobile terminals based on their decision mechanisms.

4 Performance Analysis

The main focus of this section is to examine the behaviour of the different types of RM architectures, and to justify the benefit of implementing IEEE 802.21 in terms of energy efficiency of the mobile terminals in a HWN environment. The simulation analysis due to simplification considers only two types of wireless technologies, WiMAX and WiFi. More complex HWN structures incorporating UMTS and Satellite networks are also under consideration.

Simulation platform. The topology of the simulated scenario is given in Figure 1, while the simulation parameters are given in Table 1. The mobile users are randomly distributed throughout the simulation area and represent multi-interface (WiFi/WiMAX) terminals.

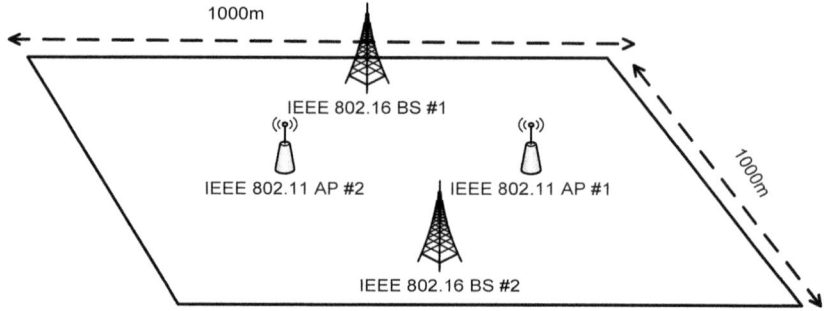

Figure 1 Simulation topology.

Table 1 Simulation parameters.

Scenario parameters	Parameter values
Simulation Duration	5 min
WiFi Standard	IEEE 802.11g
WiMAX Standard	IEEE 802.16e
WiFi AP antenna gain	15 dB
WiFi AP antenna height	5 m
WiMAX BS antenna gain	18 dB
WiMAX BS antenna height	15 m
Shadowing	4 dB
Application Type	CBR/UDP
Battery Capacity	1200 mAh
Battery model	Linear
Mobility model	Random way point
Number of mobile nodes in the scenario	30, 90

The analysis is carried out in the QualNet [17] network simulator that has a specific ability to calculate the energy consumption of the mobile nodes in the simulated scenario. The IEEE 802.21 module and the multi parametric RM are additionally developed and implemented as standalone external applications that communicate with the simulator via TCP sockets [16, 18], while the SNR based RM is implemented in the simulator.

The metric used for the simulation analysis is the *energy consumption* of the mobile terminals. It is defined as a sum of the consumed energy (in Joules) of the *receive*, *transmit* and *idle* state of the mobile terminals, and it is extracted from the battery level condition at the end of the simulation. The results depict the *energy consumption* in dependence of the RM features that were discussed in the previous section: *application bitrate*, *mobile terminal speed*

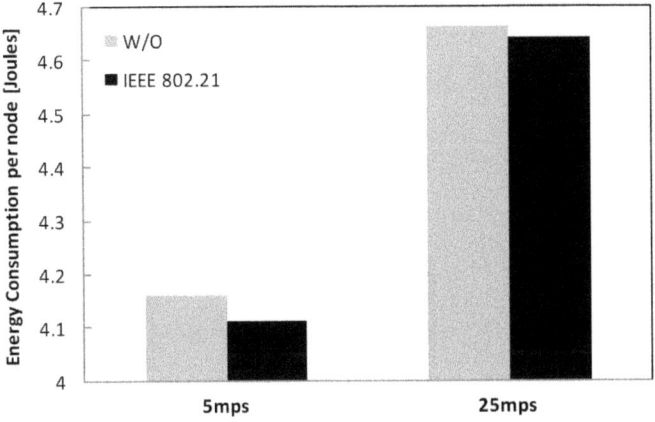

Figure 2 Energy consumption for different speeds of the mobile terminals.

and *number of users*. The energy consumption in the results is calculated as an average of the consumed energy per node in the simulation scenario.

4.1 IEEE 802.21 Benefits

This subsection analyzes the impact of the IEEE 802.21 implementation in mobile terminals on the energy consumption. The energy consumption for different speeds of the mobile terminal, with and without IEEE 802.21 implementation, is given in Figure 2. There are 90 mobile nodes and they use a 64 kbps CBR/UDP type of application in the simulation scenario. The lower speed on the figure (5 mps) can be related to the speed of the mobile terminals in an urban surrounding, while the higher speed (25 mps) is relevant for mobile terminals moving on highways. The results obtained for the case when the nodes are using IEEE 802.21 are compared with the results obtained for the case when the mobile nodes do not use IEEE 802.21. It is evident that when using IEEE 802.21 less energy is consumed in the mobile terminals. IEEE 802.21 achieves larger energy efficiency for lower speeds of the mobile terminals. The increased mobile terminal speed results in an increased number of horizontal and vertical handovers. The horizontal handovers have a negative influence on the energy efficiency gain of the IEEE 802.21 standard, hence increased number of horizontal handovers result in decreased gain in terms of the energy efficiency.

The energy consumption for different application bitrates, with and without IEEE 802.21 implementation, is depicted in Figure 3. The speed of

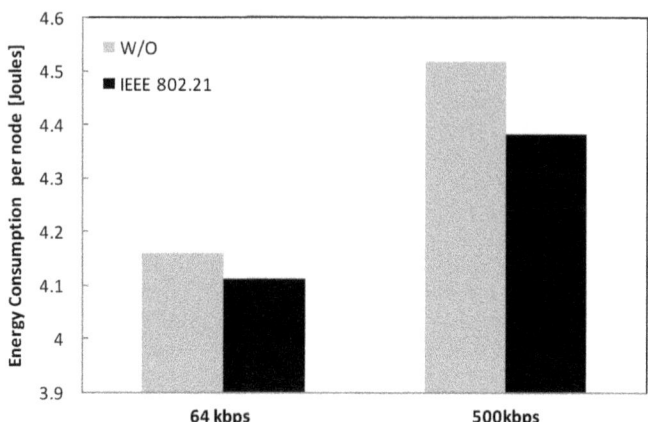

Figure 3 Energy consumption for different application bitrates.

the mobile nodes is 5 mps. The lower bitrate (64 kbps) in the given scenario emulates a voice session while the higher one (500 kbps) emulates a video session. It is evident that IEEE 802.21 enables more energy efficient communication for both types of applications. It is also noticeable that larger benefit (larger energy efficiency) is obtained when using IEEE 802.21 for high applications bitrates. For high application bitrates and large number of users in the network the mobile terminals will rarely change the point of attachment in order to preserve the needed QoS level. Hence smaller amount of horizontal handover will be performed resulting in an increased energy efficiency gain of the IEEE 802.21 standard.

The energy consumption for different number of mobile nodes is depicted in Figure 4. As seen from the figure, IEEE 802.21 has better performance for any number of nodes in the scenario with the benefits of implementing IEEE 802.21 being almost identical for both small and large number of nodes.

The results show that IEEE 802.21 lowers the consumed energy in the mobile terminals, thus justifying the benefit of its implementation in terms of energy efficiency. The next subsection will focus on the energy consumption performance between the single parametric RM and multi parametric RM.

4.2 Resource Management Benefits

This subsection analyzes the impact of the two RM architectures elaborated in Section 3.2 on the energy consumption in the mobile terminals. The multi-parametric RM is compared to the single-parametric one (SNR based)

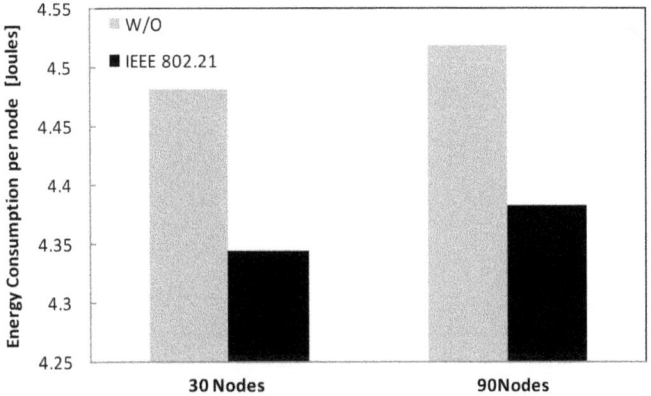

Figure 4 Energy consumption for different numbers of nodes.

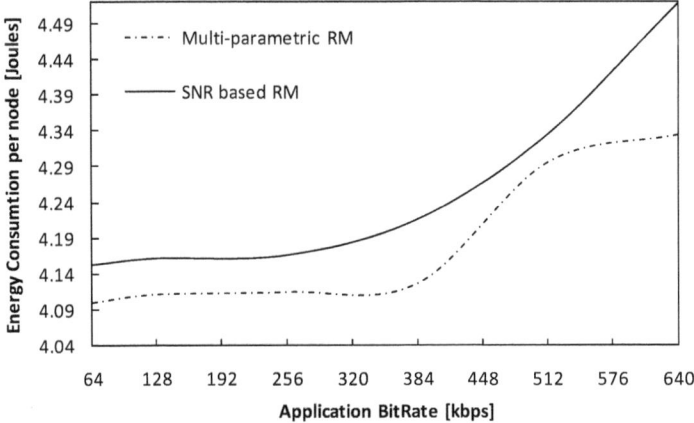

Figure 5 Energy consumption in dependence of the application bitrate.

as a most common RM architecture. Both architectures are simulated with the IEEE 802.21 standard implemented in order to provide the interoperability features. The energy consumption is analyzed in dependence of the application bitrate and mobile terminal speed.

The energy consumption in dependence of the application bitrate for both types of RM architectures is depicted in Figure 5. It is evident that the multi-parametric RM has lower energy consumption than the SNR based RM. It must be stressed that in the given analysis the multi-parametric RM has not been designed to take the energy consumption into consideration in the net-

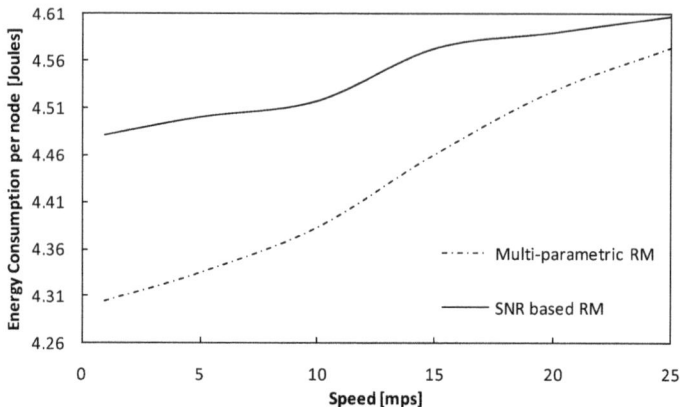

Figure 6 Energy consumption in dependence of the mobile terminal speed.

work selection process. The multi-parametric RM selects the most suitable network based on multiple input parameters, achieving a more optimal decision in comparison to the single parametric decision of the SNR based RM, which results in lower energy consumption.

The energy consumption of both RM architectures in dependence of the mobile terminal speed is given in Figure 6. It is obvious that the multi-parametric RM consumes lower amount of energy than the SNR based RM. For lower mobile terminal speeds, the difference in the energy consumption between both RM architectures is high. As the mobile speed increases, the gain of the multi-parametric RM decreases due to the frequent initialization of the handover procedures. The results in Figure 7 are obtained as an average from a vast set of simulations executed for different values of the input parameters reflecting the influence of the application bitrate and the mobile terminal speed on the energy consumption for both RM architectures. It can be concluded that the multi-parameter RM outperforms the SNR based RM in terms of the energy consumption in all cases.

In the previous figures, the multi-parametric RM decision process did not take the energy consumption into consideration as a separate optimization parameter. Figure 8 depicts the consumed energy in dependence of the application bitrate for a multi-parametric RM [15] and its modified, energy aware, version (both based on neural networks mechanisms). The figure clearly shows that introducing the energy awareness in the RM architectures decreases the energy consumption thus better performance can be obtained. It must be stressed that the energy efficiency gain in Figure 8 can be ad-

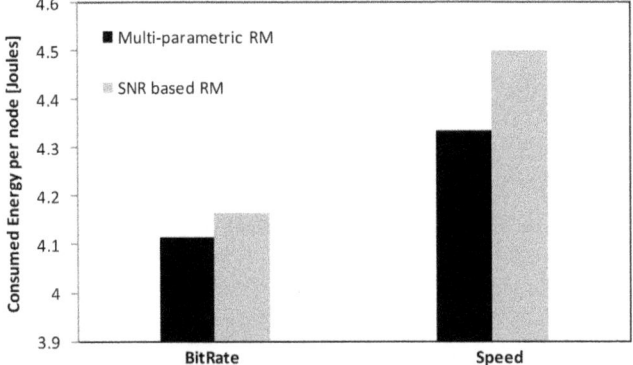

Figure 7 General representation of the energy consumption.

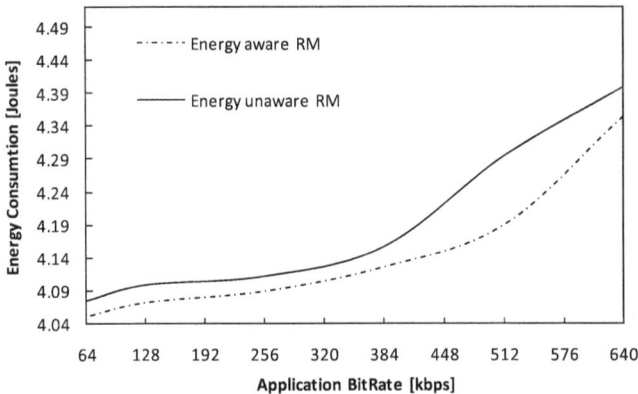

Figure 8 Energy consumption for an energy aware and unaware RM architecture.

ditionally improved by increasing the weight of the energy awareness in the optimization performed by the multi-parametric RM decision in HWN context.

5 Conclusions

A key entity in the Heterogeneous Wireless Networks is the Resource Manager. Its role is to ensure service continuity while preserving the needed QoS level to the end users. The State of the Art design of the mobile terminals leads to an increase of the number of physical interfaces which directly affects the energy consumption of the terminals. This fact implies that the

RM modules should be also designed in a manner that will enable an energy efficient communication of the mobile users. This paper analyzes the benefits of implementing the IEEE 802.21 protocol and examines two different RM architectures in terms of the energy consumption. It is evident that implementing IEEE 802.21 enables a more energy efficient communication. Also, the results show that the multi-parametric RM architecture has lower energy consumption rate in comparison to the single parameter RM architectures. This behavior is a result of the superior multi parameter network selection in comparison to the single parameter network selection. As seen from the figures, suboptimal network selection can lead to increased energy consumption due to frequent handover initialization, bad channel state, network overload, frequent retransmissions etc. Moreover, additional energy efficiency can be achieved if the energy awareness is introduced in the multi-parametric RM. The RM should be capable of choosing the most energy efficient network while preserving the needed QoS level of the mobile users.

Future work will focus on further development of the multi-parametric RM in terms of the energy efficiency, and analyzing its performance throughout complex simulation scenarios.

References

[1] L. Gavrilovska and R. Prasad. *Ad Hoc Networking towards Seamless Communications*, Springer, 2006.
[2] E. Hossain (Ed.). *Heterogeneous Wireless Access Networks: Architectures and Protocols*, Springer, 2009.
[3] IEEE 802.21: Media Independent Handover. Information available at http://www.ieee802.org/21.
[4] Huaiyu Liu Maciocco, C. Kesavan, and V. Low. Energy efficient network selection and seamless handovers in Mixed Networks. In *Proceedings of IEEE International Symposium on a World of Wireless, Mobile and Multimedia Networks & Workshops*, 2009.
[5] J.-H. Yeh, J.-C. Chen, and C.-C. Lee. Comparative analysis of energy-saving techniques in 3gpp and 3gp2 systems. *IEEE Transactions on Vehicular Technology*, 58:432–448, 2009.
[6] N. Balasubramanian, A. Balasubramanian, and A. Venkataramani. Energy consumption in mobile phones: A measurement study and implications for network applications. In *Proceedings of the 9th ACM SIGCOMM Conference on Internet Measurement*, pp. 280–293, 2009.
[7] M. Nillson and L.M. Feeney. Investigating the energy consumption of a wireless network interface in an ad hoc networking environment. In *Proceedings of the Twentieth Annual Joint Conference of the IEEE Computer and Communications Societies (INFOCOM'01)*, Vol. 3, pp. 1548–1557, 2001.

[8] SuKyoung Lee and N. Golmie. Power-efficient interface selection scheme using paging of WWAN for WLAN in heterogeneous wireless networks. In *Proceedings of IEEE International Conference on Communications (ICC'06)*, pp. 1742–1747, 2006.
[9] C. Desset, N. Ahmed, and A. Dejonghe. Energy savings for wireless terminals through smart vertical handover. In *Proceedings of IEEE International Conference on Communications*, 2009.
[10] S. Minji, S. Nakjung, and S. Yongho. WISE: Energy-efficient interface selection on vertical handoff between 3G networks and WLANs. In *Proceedings of 15th IEEE International Symposium on Personal, Indoor and Mobile Radio Communications*, Vol. 1, pp. 692–698, 2004.
[11] D.N. Zuckerman. Green communications – Management included. In *Proceedings of IEEE ICC09 workshop on Green Communications*, Dresden, Germany, June 2009.
[12] V. Rakovic, V. Atanasovski, and L. Gavrilovska. Velocity aware vertical handovers. In *Proceedings of 2nd International Symposium on Applied Sciences inBiomedical and Communication Technologies (ISABEL09)*, 2009.
[13] Advanced Resource Management Solutions for Future All IP Heterogeneous Mobile Radio Environments (AROMA) project, http://www.aroma-ist.upc.edu, IST-4-207567.
[14] X. Gelabert et al. On the suitability of load balancing principles in heterogeneous wireless access networks. In *Proceedings of IWS 2005/WPMC'05*, Aalborg, 2005.
[15] V. Rakovic and L. Gavrilovska. Vertical handover mechanism for future heterogeneous wireless systems. In *Proceedings of IEEE International Congress on Ultra Modern Telecommunications and Control Systems (ICUMT 2010)*, Moscow, 2010.
[16] V. Atanasovski, V. Rakovic, and L. Gavrilovska. Efficient resource management in future heterogeneous wireless networks: The RIWCoS approach. In *Proceedings of IEEE Military Communications Conference (MILCOM 2010)*, San Jose, CA, USA, October 31–November 3, pages 1592–1597, 2010.
[17] QualNet. Available at http://www.scalable-networks.com/products/developer.php.
[18] P. Latkoski, V. Rakovic, O. Ognenoski, V. Atanasovski, and L. Gavrilovska. SDL+QualNet: A novel simulation environment for wireless heterogeneous networks. In *Proceedings of 3rd International ICST Conference on Simulation Tools and Techniques*, 2010.

Biographies

Liljana Gavrilovska currently holds the positions of full professor, Head of the Institute of Telecommunications and the Center for Wireless and Mobile Communications (CWMC) at the Faculty of Electrical Engineering and Information Technologies, Ss Cyril and Methodius University (UKIM) in Skopje. She received her BSc, MSc and PhD from UKIM in Skopje, University of Belgrade and UKIM in Skopje, respectively. Dr. Gavrilovska is author/co-author of more than 100 research journal and conference publications and technical papers, co-author of the book *Ad Hoc Networking Towards Seamless Communications* (Springer, 2006) and co-editor of the book *Application and Multidisciplinary Aspects of Wireless Sensor Networks* (Springer, 2010). Her major research interest is concentrated on cognitive radio networks, future mobile systems, wireless and personal area networks, cross-layer optimizations, ad-hoc networking, traffic analysis and heterogeneous wireless networks. She is a senior member of IEEE and serves as a Chair of the Macedonian ComSoc Chapter.

Valentin Rakovic currently holds the position of teaching and research assistant at the Institute of Telecommunications at the Faculty of Electrical Engineering and Information Technologies, Ss Cyril and Methodius University in Skopje. He has received his BSc and MSc from Ss Cyril and Methodius University in Skopje, in 2008 and 2010, respectively. He is currently involved in the FP7 funded QUASAR, FARAMIR and ACROPOLIS projects, and has also participated in a NATO funded SfP project – RIWCoS. Valentin Rakovic is an author/co-author of around 20 research journal articles, conference publications and technical papers. His major research interests lie in the areas of cognitive radio networks, resource management for heterogeneous wireless networks, artificial intelligence algorithms.

Vladimir Atanasovski currently holds the position of assistant professor at the Institute of Telecommunications at the Faculty of Electrical Engineering and Information Technologies, Ss Cyril and Methodius University (UKIM) in Skopje. He received his BSc, MSc and PhD from UKIM in Skopje, in 2004, 2006 and 2010, respectively. Dr. Atanasovski is an author/co-author of more than 50 research journal articles, conference publications and technical papers. His major research interests lie in the areas of cognitive radio networks, resource management for heterogeneous wireless networks, traffic analysis and modeling, cross-layer optimizations and ad-hoc networking.

A Cloud Infrastructure to Manage Future Internet: The Virtual Network Operation Center

P. Donadio[1], A. Cimmino[1] and R. Prasad[2]

[1] Optical Network Division (OND), Alcatel Lucent Italy, 84091 Salerno, Italy;
e-mail: {pasquale.donadio, antonio.cimmino}@alcatel-lucent.it
[2] Department of Electronic Systems, Aalborg University, 9220 Aalborg, Denmark;
e-mail: prasad@es.aau.dk

Received: 31 January 2011; Accepted: 15 March 2011

Abstract

A Network Operations Center (NOC) is a place from which operators and administrators supervise, monitor and maintain a converged telecommunications networks. The NOC is the focal point for network troubleshooting, software distribution and updating, router and domain name management, performance monitoring, and coordination with affiliated networks. This paper analyzes and proposes a set of innovations for current NOC infrastructure when the NOC is virtualized, using IaaS[1] and PaaS[2] cloud-aware facilities. The scenario described will show how the setting up and operation of the V-NOC should be enabled by emerging technologies for mobile and fixed network business and related technologies.

Keywords: cloud computing, virtualization, network management, next generation networks, Service Oriented Architecture (SOA), Service Oriented Infrastructure (SOI).

---

[1] Infrastructure as a Service.
[2] Platform as a Service.

Journal of Green Engineering, 255–265.
© 2011 *River Publishers. All rights reserved.*

1 Introduction

With the rapid development of cloud technologies and their popular applications, leading to the more and more exacting problems of complex computations and massive data process, arguments for high performance processing devices are becoming increasingly vehement. Nowadays, the number of Intranet formed of many computer clusters is quickly increasing, while cheap personal computers are distributed everywhere, with a low rate of resource usage however. Cloud computing [1, 2] techniques as well as pervasive computation schemes may become the suitable and cost effective approaches to take concrete advantage of the underutilized resources.

The current evolution pushes the redesign of the Network Management architecture to a complex cloud-based purposive infrastructure. Virtualized multiple-agent technology [3, 4] represents an exciting new perspective of analyzing, designing and building complex network management systems. The autonomous, cooperative as well as purposive infrastructure [3] with intelligent features of an agent make explicit that the agent-based system becomes a promising software solution in virtualized cloud computing environments [5].

Within the next generation Network Management System (NMS), network resource are structured via XaaS [6] (IaaS, SaaS, and so on) [14]. All entities and resources involved in the cloud are viewed as services. The SOA [6] accelerates the convergence of agents and cloud. One of the challenging issues in service-oriented cloud using agent-based systems is how to facilitate the service composition with reference to the dynamic feature of the cloud environment. Virtualization orientation is an increasingly recognized paradigm for agent modeling and development.

In this paper, we present an innovative approach for modeling, designing, and managing agent mediated cloud services for TMN [7] application framework. The Virtual Network Operation Center model is introduced for modeling and designing various service agents in a concrete telecommunication scenario. Based on the virtualization approach and cloud-based multi-agent system, a new infrastructure is proposed to enable various network management service functions such as service advertisement, service discovery, service negotiation, and service delivery functions in service-oriented cloud environment. Within the cloud modeling approach, service agents are proposed for the management of the distributed resources, discovery and selection of computing services, etc. The service agents are also used to enable the autonomous behavior of the system, i.e. to adapt to users' computation needs and dynamic resource environments. This paper is organized as follows: Section 2 describes the background and related works about cloud computing and virtualization in a network management scenario. Section 3 introduces the virtual network operation center architecture. It also illustrates the basic requirements of the proposed architecture. Section 4 presents the V-NOC workflow. Finally, a conclusion and the future work are discussed in Section 5.

2 Background and Related Works

According to the V-NOC architecture there are five components (Fault, Configuration, Accounting, Performance and Security management – FCAPS [8]) involved in network management and three components used for service management (Monitoring, Control, Reporting).

The Virtual Network Operations Center (V-NOC) covers all components with a combination of open source tools or instruments that are presented in detail in the following section.

Fault management has to do with network problems discovery and correction. Potential problems are identified, and steps are taken to prevent them from occurring or recurring. This way, the network is kept operational and downtime is minimized. The correction of discovered problems is not automatic, but rather follows a path of procedures and communication between NMS, Helpdesk and the Operator.

Configuration management is responsible for network operation control. Hardware and programming changes, including the addition of new equipment and programs, modification of existing systems and removal of obsolete systems and programs, are coordinated. An inventory of equipment and programs is kept and updated regularly.

Accounting management is devoted to distributing resources optimally and fairly among network subscribers. This makes the most effective use of the systems available, minimizing the cost of operation. This level is also responsible for ensuring that users are billed appropriately.

Performance management is involved with managing the overall performance of the network. Throughput is maximized, bottlenecks and other potential problems are identified. A major part of the effort is to identify which improvements will yield the greatest overall performance enhancement.

At the *Security management* level, the network is protected against offenders and denial-of-service attack (DoS attack), unauthorized users, and physical or electronic sabotage. Confidentiality of users' information is maintained where necessary or warranted. The security systems also allow network administrators to control what each individual authorized user can (and cannot) do with the network equipment.

Monitoring of services involves gathering data about the network services. The following services are monitored: status of interfaces on border routers, status of BGP sessions and the size of the routing table, CPU utilization on routers, MPLS status.

Control refers to manipulation of devices. No automatic manipulation is planned for the first burst of planned operations; rather, all such intervention will be accomplished by human interaction.

Reporting refers to documenting abnormal events and circulation of these documents. It will be materialized by the Helpdesk and the TTS (Trouble Tickets System).

Providing hybrid, agent-based operational cloud-like services architecture is still challenging. A number of initiatives and funded projects for applying cloud based virtualized agents have appeared in recent years. RESERVOIR [9, 10] focuses on technologies that enable to build cooperating computing clouds in order to connect computing clouds to create an even bigger cloud. The Service Oriented Infrastructure (SOI), the resource sharing across organizations/geographies and the use of virtual machines as the basic unit of work, are the key issues of the project.

The idea of GEYSERS on the other hand [11] is the convergence of IT and Telco infrastructure service provisioning, control and management to deploy new cloud scenarios to reach a new level of critical mass. This convergence could support greater flexibility and efficiency in the way IT departments operate and coordinate the provisioning of all of these resources across the cloud, but also how they enact outsourcing of IT capability maintaining essential cloud characteristics: protocol transparency, redundancy options, space and power efficiency. This project proposes a virtual infrastructure layer to handle most of the infrastructure virtualized resources and represent a kind of mediator, virtual infrastructure providers, between the owner of the infrastructures and virtual Infrastructure operators.

The rationale behind the proposed paper is to use a mixed cloud architecture (IT and Network) to support a Virtual Operation Center that could be organized, configured and deployed in a dynamic way, reducing the OPEX and CAPEX of the overall telecommunication infrastructure.

3 The V-NOC Architecture

This section describes the architecture of the V-NOC taking into account current requirements and possible future extensions. The architecture design was mainly driven by the following basic *requirements*:

- *Scalability.* The principal reason for having a distributed infrastructure instead of a centralized one is the higher scalability of decentralized solutions. The V-NOC architecture has to be able to support lots of users in the near future, sharing their knowledge and data, currently only accessible via the local desktops of the NOC. Whereas a centralized solution is only able to support a limited number of users, scalability is a major decentralization design requirement even though it is sometimes in conflict with other requirements for the architecture.
- *Fault-tolerance.* A decentralized system consisting of unreliable loosely coupled nodes (e.g., user's desktops and laptops) have to be able to deal with failures such as network churn or node failures. The architecture has to take this into account, up to a certain degree, the system availability without any loss of service quality.
- *Flexibility.* The V-NOC system has to be able to support future extensions by new technologies and Telco service distribution, such as provider/consumer/prosumer paradigms. A flexible architecture is required to be able to integrate those extensions and offers new functionalities through existing APIs. The V-NOC is independent from the underlined multi-vendor technology and from the specific O&M system available normally at layer level or deployed nodes level. More in general the network management layer applied to this virtualization of resources shall therefore allow configuring the dynamic downgrade or upgrade of available infrastructure resources.

Apart from new technologies, the infrastructure also has to take into account domain specific requirements variation, i.e., users have to be able to tailor the V-NOC system to their needs by providing application-specific handlers.

In addition to the requirements shown above, basic overall purposes of the V-NOC architecture are:

- To provide a V-NOC to third part entities (network providers, operators, vendors etc.) in order to maximize the network management business (e.g. outsourcing) and optimize the resource usage.

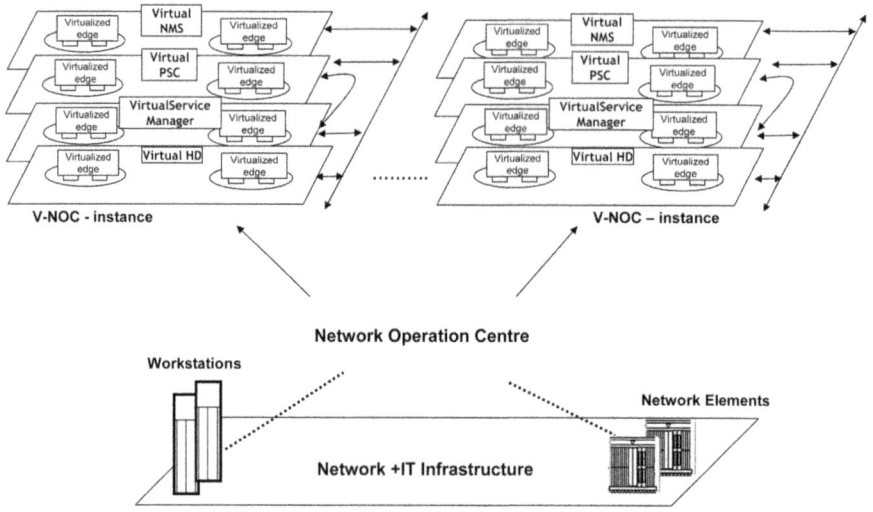

Figure 1 The V-NOC architecture.

- To provide an (all-in-one) immersive environment with familiar NOC structures such as Network Elements, distributed storage systems, power equipment, and displays.
- To provide a multi-user virtual world where users can effectively collaborate on elements of the NOC together.
- To provide a centralized graphical tool useful to manage network resources as a real NOC, but which can be also used as a modeling and simulation tool (e.g. requirements monitoring and analysis) in order to make better use of network and storage resources, discussing measurements and consolidate monitoring.

The V-NOC is organized as an IaaS scheme, as presented in Figure 1. In the V-NOC concept the network operations and services management are provided by different entities in a distributed paradigm. Each instance of the V-NOC is composed of the following components:

- *Virtual Network Management System* (V-NMS): a small group of network managers formed by access port managers that have the administrative control of the overall Network Elements connected to the V-NOC, guaranteeing proper functioning. NMS provides network management and user support within their area of authority.
- *Virtual Help Desk*: This entity is entitles to monitor connectivity problems and to handle the trouble tickets System.

- *Virtual Service Manager*: These are a set of virtual entities which design, specify and orchestrate the deployment of advanced services on the V-NOC infrastructure (networking and storage services).
- *Virtual Project Steering Committee* (PSC): It is not considered part of the V-NOC, however it interacts with the access port managers and it makes decisions on strategic aspects of the project, and the deployed services.

4 The V-NOC Workflow

The V-NOC workflow will be organized in five different phases:
- Phase 1: The V-NOC is managed virtually. Basic components of the V-NOC are accessible in ubiquitous way.
- Phase 2: Part of the V-NOC is the Virtual Network Management System. It is composed of enhanced graphical users interfaces in order to plan, provision and monitor (in real-time) the physical changes within the Infrastructure Provider's network.
- Phase 3: New virtual services are offered using the Virtual Service Manager. These services are used to design, specify and orchestrate the deployment of advanced services on the V-NOC infrastructure.
- Phase 4: When a problem occurs it is possible to interact with the Virtual Help Desk system, in order to solve connectivity problems and handles the trouble tickets System.
- Phase 5: The Virtual PSC is available to makes decisions on strategic aspects of the project and the corresponding deployed services.

5 Conclusions and Future Work

In this paper, we have described the evolution of the Network Operation Centre (NOC) architecture [19] by using virtualization over the cloud. We have discussed a Virtual Network Operation Centre (V-NOC) that provides the runtime machinery to easily manage next generation networks, using a distributed approach.

Its integration into the global cross-platform NMS has been made possible via support for execution of virtualized jobs through cloud interface using a broker middleware (open-nebula). We plan to extend the V-NOC to a wide variety of services: (1) The support of additional functionalities via the API including inter-thread communication is planned. (2) We are

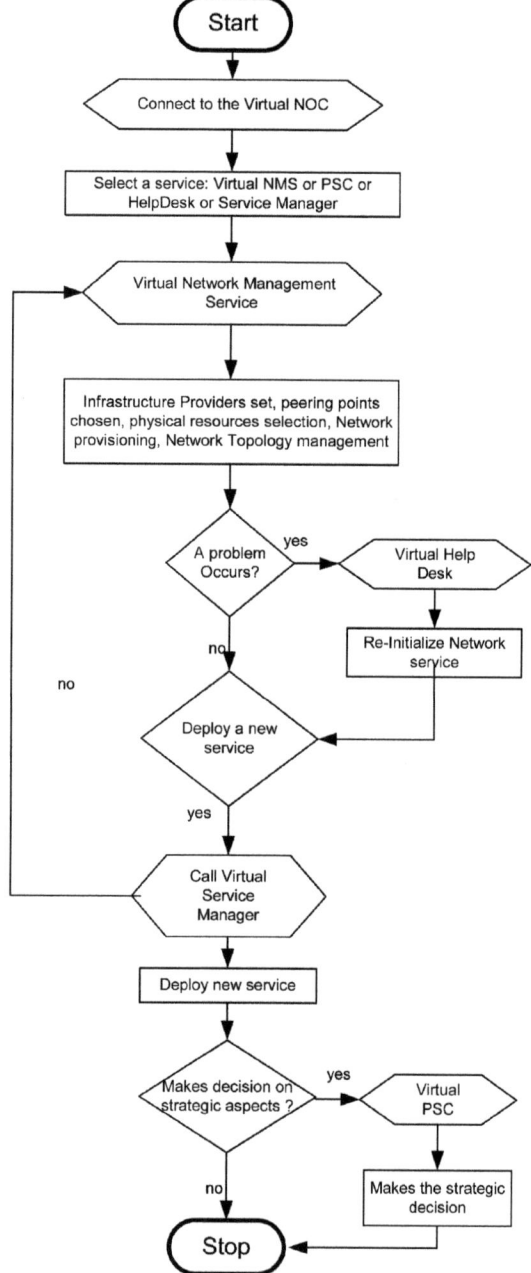

Figure 2 The V-NOC workflow.

working on support for multi-clustering with Peer-to-Peer communication between Network Managers. (3) We plan to support utility-based resource allocation policies driven by economic, quality of services, and service-level agreements. (4) Since Internet has been largely deployed and utilised around the globe several business models have been adopted to yield revenues to all actors involved in the value chain. A consolidated one is based on gold services associated with "free" basic service offers. Good examples are YouTube, megaupload, megavideo, Skype and recently announced Google on net QoS.

Therefore, on top of the best effort quality, the internet is converging to a quality managed cloud that shall need both high capacity generalised infrastructure and innovative generalised OSS/BSS (V-NOC) solutions also exploiting APIs to the service providers like GEYSERS concept.

We are also investigating strategies for adherence to Web Service Resource Framework (WSRF) and Representational State Transfer (REST) standards by extending the current V-NOC interface. This is likely to be achieved by its integration with open-nebula [12] low-level cloud middleware implementations that conform to OCCI standards such as OGF [13]. Finally, we plan to provide data cloud capabilities to enable resource providers to share their data resources in addition to computational resources.

Acknowledgements

The authors would like to thank the FP7-GEYSERS (FP7-ICT-248657) project staff for their support and contributions.

References

[1] Cloud Computing, http://en.wikipedia.org/wiki/cloud_computing.
[2] David Patterson. Cloud computing and the RAD LAB, UC Berkley. Keynote lecture presented at Cloud Futures 2010, http://perspectives.mvdirona.com/2010/05/04/PattersonOnCloudComputing.aspx.
[3] Multi-Agent System, http://en.wikipedia.org/wiki/multi-agent_system.
[4] H. Arafat Ali. Multi-agent system for specific domain search engine based on distributed classification approach. Computers Engineering & Systems Department, Faculty of Engineering, Mansoura University of Mansoura, Egypt.
[5] Cloud Computing and Virtualization, http://www.cloudcomputingconf.org/.
[6] XaaS – Anything As a Service, http://www.xaas.com/.
[7] TMN: Telecommunications Management Network Standards, www.itu.int/TMN [M.3010–M.3400].

[8] FCAPS: Fault, Configuration, Accounting, Performance, and Security Standards, http://www.iso.org.
[9] B. Rochwerger, J. Caceres, R.S. Montero, D. Breitgand, E. Elmroth, A. Galis, E. Levy, I.M. Llorente, K. Nagin, and Y. Wolfsthal. The RESERVOIR model and architecture for open federated cloud computing. *IBM Systems Journal*, 53:4, 2009.
[10] B. Sotomayor, R.S. Montero, I.M. Llorente, and I. Foster. Virtual infrastructure management in private and hybrid clouds. *IEEE Internet Computing*, 13(5), September/October 2009.
[11] GEYSERS: Generalised Architecture for Dynamic Infrastructure Services, http://www.geysers.eu/.
[12] Open Nebula, http://www.opennebula.org.
[13] OGF Open Cloud Computing Interface, http://www.occi-wg.org/doku.php.
[14] http://www.scribd.com/doc/29063258/Converged-Optical-Network-Infrastructures-in-Support-of-Future-Internet-and-Grid-Services-Using-IaaS-to-Reduce-GHG-Emissions

Biographies

Pasquale Donadio is a system engineer working for Alcatel-Lucent Italia. He received the automation engineering degree from the University of Naples Federico II, where his thesis work encompassed the modeling and simulation of innovative multimedia languages based on XML. He began his career at IPM, where he worked on the design of Internet security systems based on smartcards. Then he joined the Alcatel Italia OND (Optical Network Division) and worked on network management design and development, first as a top-level designer and later as system architect. Since his graduation he has been continuing his research activity acting in a part-time position in several Frame Program 7 (FP7) funded programs, as task leader and wp leader (ETICS, GEYSERS, ECONET, OneLAB2, E-photon/ONe+). His current research interests include network management based on distributed systems (Grid&Cloud computing), energy aware networks and tridimensional graphical user interfaces for network management and control. He has authored/co-authored about 20 patents.

Antonio Cimmino graduated in electronic engineering in Napoli (I). He is trainer at the Italian Air force in telecommunications and air-navigation systems. In 1991 he joined Alcatel Italia, working in the radio mobile research department, primarily involved in mobile research activities: system definition and network architecture on Mobile Broadband System project (MBS/60 GHz), UMTS – Monet, OBANET and Moicane (FP5). Recently, he has been involved in co-ordination of FP6/FP7 research projects (WEIRD, OneLab 1&2, ePhoton+, Nobel 2, GEYSERS and ECONET) for the IST area.

Ramjee Prasad (R) is a distinguished educator and researcher in the field of wireless information and multimedia communications. Since June 1999, Professor Prasad has been with Aalborg University (Denmark), where currently he is Director of Center for Teleinfrastruktur (CTIF, www.ctif.aau.dk), and holds the chair of wireless information and multimedia communications. He is coordinator of European Commission Sixth Framework Integrated Project MAGNET (My personal Adaptive Global NET) Beyond. He was involved in the European ACTS project FRAMES (Future Radio Wideband Multiple Access Systems) as a Delft University of Technology (the Netherlands) project leader. He is a project leader of several international, industrially funded projects. He has published over 500 technical papers, contributed to several books, and has authored, coauthored, and edited over 30 books. He has supervised over 50 PhDs and 15 PhDs are at the moment working with him. He has served as a member of the advisory and program committees of several IEEE international conferences. In addition, Professor Prasad is the coordinating editor and editor-in-chief of the Springer International Journal on *Wireless Personal Communications* and a member of the editorial board of other international journals. Professor Prasad is also the founding chairman of the European Center of Excellence in Telecommunications, known as HERMES, and now he is the Honorary Chair. He has received several international awards; the latest being the "Telenor Nordic 2005 Research Prize". He is a fellow of IET, a fellow of IETE, a senior member of IEEE, a member of The Netherlands Electronics and Radio Society (NERG), and a member of IDA (Engineering Society in Denmark). Professor Prasad is advisor to several multinational companies. In November 2010, Ramjee Prasad received knighthood from the Queen of Denmark, the title conferred on him is Riddere af Dannebrog.

Green Communications: An Emerging Challenge for Mobile Broadband Communication Networks

E. Calvanese Strinati, A. De Domenico and L. Herault

CEA, LETI, MINATEC, 38000 Grenoble, France;
e-mail: {emilio.calvanese-strinati, antonio.de-domenico, laurent.herault}@cea.fr

Received: 7 February 2011; Accepted: 20 March 2011

Abstract

Worldwide mobile broadband communications networks are increasingly contributing to global energy consumption. In this paper we tackle the important issue of enhancing the energy efficiency of cellular networks without compromising coverage and users perceived Quality of Service (QoS). The motivation is twofold. First, operators need to reduce their operational energy bill. Second, there is a request of environmental protection from governments and customers to reduce CO_2 emissions due to information and communications technology. To this end, in this paper we first present the holistic system view design adopted in EARTH (Energy Aware Radio and neTworking tecHnologies) project. The goal is to ensure that any proposed solution to improve energy efficiency does not degrade the energy efficiency or performance on any other part of the system. Then, we focus on technical solutions related to resource allocation strategies designed for increasing diversity order, robustness and effectiveness of a wireless multi-user communication system. We investigate both standalone and heterogeneous cells deployment scenarios. In standalone cells deployment scenarios, the challenge is to reduce the overall downlink energy consumption while adapting the target of spectral efficiency to the actual load of the system and meeting the QoS. Then, with heterogeneous deployment scenarios, different cell scales that ranges from macro to micro, pico and even femto cells, potentially may share the same spectrum in a given geographical area. In such scenarios interference is the most limiting problem to achieve the desired performance. Our analysis reveals how proposed methodologies permit to achieve notable energy gain over traditional resource allocation techniques especially in not saturated scenarios.

Keywords: EARTH, green communications, energy efficiency, femtocells, interference, radio resource management.

1 Introduction

Telecommunication has experienced a tremendous success causing proliferation and demand for ubiquitous heterogeneous broadband mobile wireless communications. Up to now, innovation aimed at improving wireless networks coverage and capacity while meeting the QoS requested by users admitted to the system. Nowadays, the number of mobile subscribers equals more than half the global population. Forecast on telecommunication market assumes an increase in subscribers, per subscriber's data rate, and the roll out of additional base stations for next generation mobile networks. An undesired consequence is the growth of wireless network's energy consumption that will cause an increase of the global carbon dioxide (CO_2) emissions, and impose more and more challenging operational cost for operators. Communication Energy Efficiency (EE) presents indeed an alarming bottleneck in the telecommunication growth paradigm. Motivated by this scenario, we outline the main investigation axes that may significantly improve the EE of broadband cellular networks, thereby reducing the cost and environmental impact of mobile broadband services. Today's mobile networks have a strong potential for energy savings. The design of mobile networks has until now been focused on reducing the energy consumption of terminals, whose battery power imposes stringent requirements on energy consumption. This has led to a situation where terminal energy consumption is only a fraction of the energy consumption of the mobile network. For example, NTT DOCOMO [1] has calculated that, for their 52 million subscribers in 2006, the energy consumption of their network per mobile user per day was 120 times greater than the daily energy consumption of a typical user's mobile phone. The increase in energy consumption of NTT DOCOMO's network over the years from 2002 to 2006 is shown to be directly related to the increase in number of base stations installed. Consequently, the optimization of energy-consumption of radio base stations should have a large impact on the overall EE of the network. As a matter of fact, recent increasing maturity of mobile technology in combination with the growing amount of equipment deployed each year have woken up the need to innovate in the field of energy efficient communications. Three main federative projects that face the challenge of enhancing the EE in mobile wireless broadband networks are: the core 5 Green Radio programme of mobile Virtual Center of Excellence (MVCE) [2, 3], GreenTouch [4], and EARTH [5, 6].

Energy efficient enhancement in wireless communication can be achieved only if improvements are experienced in the whole communication chain for different operational load scenarios. Several investigations are carried out in this research area, ranging from energy efficient cooling of base stations to innovative energy efficient deployment strategies, and frequency planning. The EARTH project sets a very innovative challenge: development of an holistic methodology to characterize the energy consumption of cellular networks including suitable metrics. Today, there exists no widely agreed methodology to evaluate the energy efficiency of cellular networks. Similarly, meaningful energy efficiency metrics are currently lacking. Both, in conjunction with reference scenarios, will enable a successful evaluation of the energy efficiency of networks. Such agreed methodology will be the enabler of a fair comparison of different concepts and technologies. EARTH's holistic methodology is presented in Section 2.

The objective of this paper is threefold. First, we present in detail the holistic system view proposed in the EARTH project, which may allow us to ensure that any proposed radio EE improvement does not degrade EE or performance on the terminal side or any other part of the system. To this end, in Section 2 we present and comment on a unified approach to improve the EE of the whole communication system. We focus on how energy consumption of mobile broadband networks may be significantly reduced in representative traffic scenarios. We consider LTE like systems which face a very challenging multi-user communication problem: many users in the same geographic area require high on-demand data rates in a finite bandwidth with a variety of heterogeneous services such as voice (VoIP), video, gaming, web browsing and others. We will present a holistic optimization methodology that includes *socio-economic* impact of broadband communication, definition of *reference scenarios*, and *targets and evaluation metrics*.

Second, we present and benchmark the EE effectiveness of a novel energy efficient Radio Resource Management (RRM) algorithm for OFDMA based systems. Since wireless is a shared medium, the system EE is affected not only by the single user efficiency, but also by the combination of time and frequency allocation according to the momentary instances of the frequency selective channel between the downlink base station and the receiving user equipments. Hence, in order to achieve EE, a system approach is required. In our vision, the multi-user scheduler should indeed allocate time and frequency resources to minimize the transmission energy cost while meeting QoS requirements of all active users admitted by a base station. The momentary system load plays an important role in the overall optimization design. Currently deployed base station are commonly designed so that they can accommodate the traffic demand at peak times. Nevertheless, cell traffic load notably varies during the day. Several researches pointed out that to save energy, base stations should perform a dynamic load and energy state arrangement, which balances extra load on a determined optimal set of base stations, thus maintaining minimum energy consumption. In this paper we do not propose novel dynamic sleep mode of base station components. We focus on resource allocation of active phases of base stations during downlink transmission. We propose a simple algorithm to trade off momentary spectral efficiency for downlink transmission power while meeting the QoS constraints of active users admitted in the cell. To this end, we propose a scheduling algorithm which splits the resource allocation process into four steps. In the first step we identify which entities (packets) are *rushing* and which are *not rushing*. Then in step two we deal with urgencies: we assign resources only to entities that have an high probability of missing their QoS requirements regardless to their momentary link quality and their potential to save energy. Then, if any resources (here chunks) are still unscheduled, in the third step we allocate resources to users (*non-rushing*) with highest momentary link quality, regardless to their QoS constraints. In the fourth step we perform energy efficient link adaptation to save downlink energy. We trade off throughput (lowering the transmission spectral efficiency and allocating a larger number of chunks to UEs) for downlink power by limiting the power budget on each chunk. In this way we attempt to minimize downlink transmission power over a time window, which provides significant additional flexibility to the scheduling algorithm. In addition to throughput, both latency and spectral efficiency enter in the tradeoff. In an extreme case of latency tolerance or low load scenarios, the scheduler could simply wait for the user to get close to the base station before transmitting or allocate all frequency resource to a single packet transmission with significant lowered power. We call the proposed algorithm a Green Adaptive Scheduler (GAS) [7].

Third, we propose an RRM algorithm to improve the EE of a heterogeneous cellular network where macrocells and femtocells [8] share the same spectrum in a given geographical area. In such heterogeneous deployment scenario, interference is the most limiting problem to achieve the desired performance, in terms of both spectral efficiency and energy efficiency. Macro users (M-UEs) close to femtocells experience *femto-to-macro interference* that may drastically corrupt the reliability of communications. Similarly, neighbour femtocells belonging to the same operators may also interfere with each other, thus creating *femto-to-femto interference*.

Recent economical investigations claim that femtocell deployment might reduce both the Operational Expenditure (OPEX) and Capital Expenditure (CAPEX) for cellular operators (see, for instance, [9]). A recent study [10] shows that expenses scale from $60000/year per macrocell to $200/year per femtocell. However, according the ABI Research [11, 12], by the end of 2012 more than 36 million of femtocells are expected to be sold worldwide with 150 million of customers. Thus, interdependence between energy efficiency, service constraints and deployment efficiency are not straight forward. Recently some researchers have faced these challenges. In [13], the authors carry out an energy efficiency comparison between different size cells. In [14], a theoretical study that describes the effectiveness of NodeB deployment and the system energy efficiency is presented. In [15], the authors present an energy saving procedure that allows femtocell Base Station (also named FAP or HeNB) to completely switch off its radio transmissions and associated processing when not involved in active calls. In [16], the authors cope with the problem of interference proposing a time division duplexing underlay scheme in which femtocells reuse the uplink macrocell bandwidth. In [17], an RRM algorithm that trades off frequency resource for energy is proposed to reduce the interference perceived at both macro and femto users. In [18], we come out with the design of a simple and effective RRM algorithm (RRM_{Ghost}), which permits to drastically improve the overall network performance. The proposed algorithm efficiently profit of the unusual communication context of femtocells for which locally few UEs compete for a large amount of transmission resource. In this paper, we investigate the system energy efficiency obtained if the RRM_{Ghost} algorithm is implemented at femtocells. A detailed description of the algorithm and numerical results are presented in Section 7

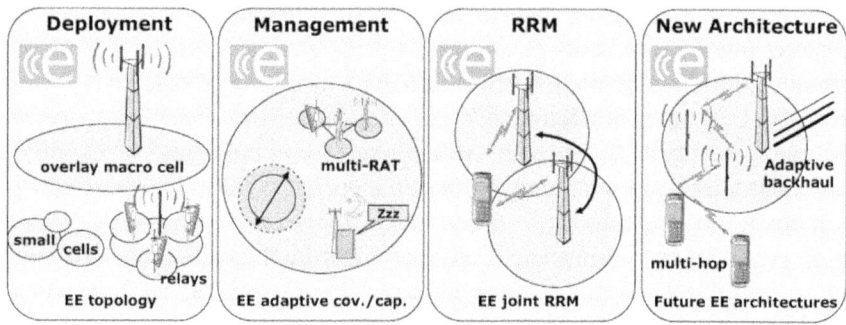

Figure 1 Green networks innovation axes.

2 Holistic EE Optimization Methodology for a Broadband Cellular Network

A unified approach to improve the EE of the whole communication system is still missing. Nevertheless, the potential offered by joint optimizations is in our vision the most promising one. Joint optimizations are represented by cross layer optimizations that take component and node architectures as well as radio interface technologies and network architecture into account. We present here the holistic optimization methodology proposed in the EARTH project that is composed of three major investigation axes:

- EE Optimization Methodology Framework
- EE Design of Green Networks
- EE Design of Green Radios

These three investigation axes will be the background knowledge for the design of new green network architecture. Figure 1 summarizes the concepts that should be investigated at network level.

3 Definition of an EE Optimization Methodology Framework

Today running broadband cellular networks have been designed to maximize coverage areas and spectral efficiency of communication systems. In such networks EE optimization has not yet played a prime role. Hence, metrics like spectral efficiency, capacity or throughput have been used to characterize usefulness and differences of investigated solutions. Nevertheless, these metrics are not sufficient to judge the EE of a communication system. A fun-

Green Communications and Mobile Broadband Communication Networks

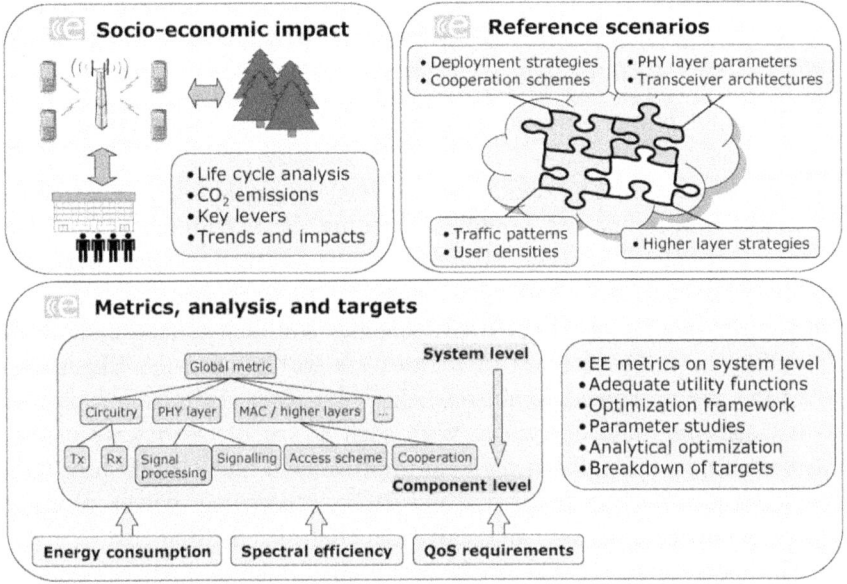

Figure 2 Framework to enhance the EE of broadband cellular networks.

damental analysis of key radio technologies in terms of their EE is lacking. This is partly due to the fact that there are neither suitable methodologies nor metrics established to enable a fair and objective comparison of these techniques. Thus, an important prerequisite for efforts to enhance the EE of cellular networks is to provide a framework to evaluate these technologies. In Figure 2 we summarize the proposed working framework.

First, the *socio-economic impact* of broadband communication system permits to define and forecast targets of energy efficient optimization. Then, the definition of *reference scenarios* characterizes the specific framework for which some specific EE enhancements are envisioned. Eventually, to evaluate the effectiveness of the EE optimization, targets and evaluation metrics are defined. Today there is not common view on such EE metrics in the wireless community and, a wide range of metrics are used by vendors operators and academia. Since 2009, standardization of energy efficiency metrics has started. One of the major bodies of standardization that is working on the topic is the ETSI Technical Committee on Environmental Engineering (ETSI-TC-EE) [19]. ETSI-TC-EE has started working on the specification of energy efficiency metrics for mobile cellular networks. Up to now, ETSI-TC-EE focuses only on EE metric specifications at component level. Furthermore, criteria related to the quality-of-service are neglected. Nevertheless, to achieve the goal of a holistic network design, EE metrics should also take the following into account:

- The input power required to generate a specific output power at the antennas.
- Energy consumed by digital signal processing equipment (interference cancelation techniques, complex decoding, etc.). With this type of metrics, the tradeoff between EE improvement of specific signal processing algorithms and their consumed energy can be understood.
- Specific metrics on EE of RRM schemes and their related control signalling cost. These types of metrics are worth when the EE analysis of the system considers not only the single-user optimization problem but also the multi-user case. With such a type of metrics we can understand the tradeoff between EE improvement due to multi-user RRM algorithms (such as frequency and packet scheduling, power allocation and power control, etc.) and additional control signalling cost.
- The energy to deliver data to the base station (backhaul power consumption).

The subject of EE metrics is presented more extensively in [20]. The EARTH European project is providing an extensive overview on this topic.

4 Design of Green Networks

Until now, several Radio Access Technologies (RATs) have been deployed in a parallel way, accommodating different service requirements over the same area and providing multi-RAT handovers. Nevertheless, such solutions have stand-alone network planning and resource management methodologies. In our method, the common network design goal has been to optimize the system's capacity and coverage while meeting QoS requirements of different types of applications in very different communication scenarios, ranging from large macro to femto cells for different environmental characteristics. Solutions have been optimized for the high load system operating point. Nevertheless, systems are most of the time offloaded. Very recently efforts are originated for joint management of the multi-RAT technologies focusing on common capacity and coverage optimization [21]. These previous efforts did not include EE considerations and not necessarily resulted in optimal deployment or performance from an energy consumption point of view. Moreover, cellular networks exhibit slowly changing daily load patterns as well as highly dynamic traffic fluctuations. However, currently networks are configured rather statically. Since most of the energy wastage occurs during low load situations, these load variations can be effectively exploited to reduce network energy consumption. Therefore, dynamic load variations should be addressed by energy efficient RRM strategies. Innovative design of green networks would require to innovate in the following three topics:

- Network Deployment.
- Network Management.
- Radio Resource Management.

4.1 Network Deployment

While a lot of work has been performed on developing throughput, capacity or coverage enhancing concepts for cellular networks, little insights have been gained which techniques can be applied to improve the energy efficiency resulting in less optimal strategies for overall network energy consumption. Moreover, recently the heterogeneous radio access networks (HetNets) concept is under investigation by both academics and industry. The HetNet deployment proposes a flexible and open architecture for a large variety of wireless access technologies, applications and services with different QoS demands as well as different protocol stacks. Wireless networks differ from each other by air interface technology, cell-size, services, price, access, coverage and ownership. The complementary characteristics offered by the different RATs make possible to exploit the deployment diversity gain leading to higher overall performance than the aggregated performances of the standalone networks. Actually, the biggest advantages of a heterogeneous network are the significant gains in network capacity and coverage via aggressive spatial spectrum reuse of available bands. This gain can be exploited to drastically reduce the overall energy consumption of the network. To this end, it is first fundamental to determine the theoretical and practical limitations on energy efficiency of HetNets deployment scenarios. The goal is to understand how to reduce system energy consumption based on deployment strategies using different cell sizes, network topologies, coordination between radio access technologies and investigates network deployments including macro base stations, femto cells, relays and repeaters that share the same spectrum in the same geographical area. Then, the second challenge will be to define how to reduce overall system energy consumption by combining and coordinating different sites and radio technologies. This is the object of EE RRM techniques. Eventually, there is the challenge of designing methodologies for energy efficient radio network deployments including femto cells, relays, repeaters, distributed antennas, etc. Indeed, an energy efficient network deployment can be design by defining the optimal mix of cell size, share of the common spectrum and, the set of adaptive resource allocation methods that may drive to large EE gains.

4.2 Network Management

In our vision, innovative network management concepts will lead to large network's energy consumption savings. The challenge is to understand how to tune the radio network nodes in order to achieve gains from coordination between different functions and nodes in the radio network, including the backbone network. The studied mechanisms should enable self EE optimization of the network, requiring only minimal human intervention. The concept is that the system autonomously and dynamically changes the cellular layout of the mobile network according to, e.g., the daily traffic profile.

To this end, the design of novel EE network management functions requires to investigate theoretical limits of energy efficient network management algorithms. Then, it is required to understand how overall system energy consumption can be reduced through proper configuration and tuning of radio network nodes, including interaction and cooperation between network elements (including radio, transport and backbone network elements). Furthermore, network management methods for EE multi-RAT coordination should be investigated.

4.3 Energy Efficient RRM

RRM algorithms have been designed to maximize the system capacity while overcoming the mismatch between requested QoS and limited network resources under full system load. However, system load in mobile systems is dynamic in nature and, traditional RRM schemes are not necessarily efficient at different operating conditions. Novel RRM schemes developed, in addition to load variation, should take into account the characteristics of RF front-end such as Power Amplifiers (PA) as well as uplink and downlink power and bandwidth constraints. This is further enhanced by the development of RRM schemes considering multi-cell cooperation for power control and scheduling transmissions. Moreover, schedulers could attempt to minimize the energy required per correctly delivered information bits, while meeting the QoS of admitted heterogeneous active users. Information theorists have studied energy-efficient transmission for at least two decades [22,23]. More recently, several researchers have designed solutions to trade spectral and energy efficiency. Single user EE optimization is a vivid investigation topic. In [24] the authors propose to trade off energy for transmission delay. Actually, Shannon theory indicates that it is desirable to transmit a packet over a longer period of time to save transmit energy. The authors consider a single user scenario with single type homogeneous traffic and, propose to scale the modulation down to conserve energy when its buffer is under loaded. When the buffer starts to fill up, the modulation order is increased to avoid long queuing times or buffer overflow. This proposal assumes that power consumption monotonically increases when transmission power increases. Under this assumption, as far as single user QoS constraints are met, power can be minimized by reducing the information rate at the expense of delay. Actually, as wireless is a shared medium, the overall system EE is affected not only by the single user efficiency, but also by the combination of time and frequency allocation according to the momentary instances of the frequency selective channel between the downlink base station and the receiving selected user equipments. Hence, in order to achieve EE, a system approach is required. In our vision, RRM methods such as the multi-user scheduler should indeed allocate time and frequency resources to minimize the transmission energy cost while meeting QoS requirements of all active users admitted by a base station. Up to now this is not the current optimization target. Traditional packet scheduling algorithms are designed to increase the maximum system capacity, subject to QoS constraints and fairness. Nevertheless, most of the time, wireless systems are only moderately loaded. With state of the art priority scheduling algorithms, the exploitation of available time and frequency resources is typically not optimized from an energy perspective.

Earliest Deadline First (EDF) [25] schedulers do not profit much from time diversity as much as they should do. Maximum Channel to Interference ratio (MCI) [26] schedulers aim at maximizing the cell throughput regardless of the user QoS and the actual system load, and consequently they are totally insensitive to any time constraints of the data traffic. In not saturated system load scenarios, the exploitation of available time and frequency resources is typically not optimized from a energy perspective. Several researches pointed out that to save energy, base stations should perform a dynamic load and energy state arrangement, which balances extra load on a determined optimal set of base stations, thus maintaining minimum energy consumption. In [7] a simple algorithm is presented to trade off momentary spectral efficiency for downlink transmission power while meeting the QoS constraints of active users admitted in the cell. The authors propose a novel scheduler, the GAS This is one example of what we can do if we change the optimization paradigm from optimization of spectral efficiency to a balanced tradeoff between energy efficiency and spectral efficiency. Moreover, in case of a multi-user scenario, power consumption of base station can be rate-independent if the system is not heavily loaded. Actually, information rate depends on transmission power but also on the momentary link quality on frequency resources allocated to scheduled users. In [27], it is recalled that the lowest order modulation should always be used while accommodating the delay constraint to minimize energy consumption.

Concluding, EE for RRM algorithms is a vivid research area. Nevertheless there is still a large room for improvement. In our vision it is required to the theoretical limits of energy efficient Radio Resource Management (RRM) algorithms and identification of their main parameters. The fundamental EE versus spectral efficiency tradeoff has not been solved yet. Furthermore, with the introduction of HetNets, an additional dimension to the tradeoff has to be considered: deployment efficiency. Finally, it is necessary to understand how highly dynamic RRM principles can reduce overall system energy consumption in practical systems where signalling overhead can drive to significant EE gain limitations.

5 Design of Green Radios

Additionally, novel technologies and components, as well as energy efficient radio interface techniques must be holistically investigated. Innovative design of green radios would require innovation of the following three topics:

- Reduction of dissipated power of base station hardware components.
- Design of link interfaces and application of innovative radio transmission techniques.
- Energy efficient RRM.

Figure 3 summarizes the concepts that we propose to investigate at node and single cell levels.

5.1 Reduction of Dissipated Power of Base Station Hardware Components

In a green network, depending on the coverage range and momentary utilization ratio of base stations, different peak output power levels will be

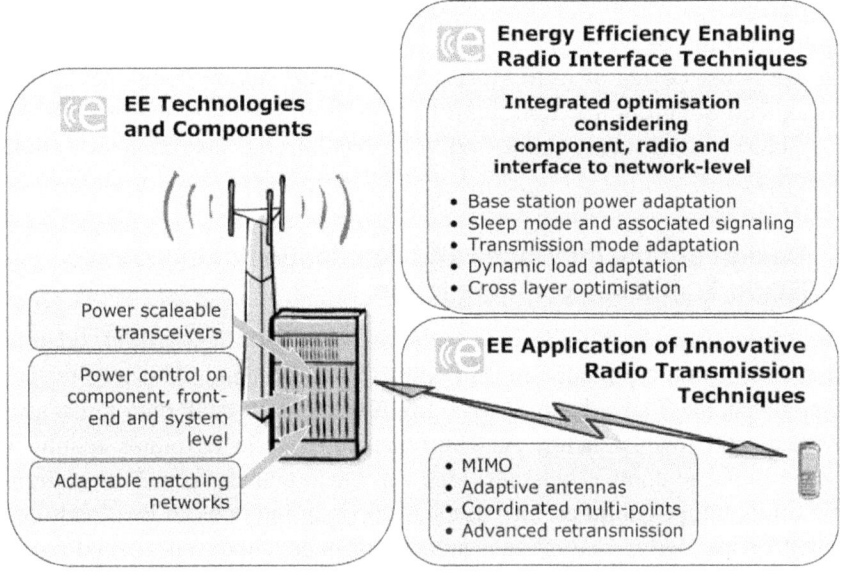

Figure 3 Green radios innovation axes.

required. State of the art Power Amplifier (PA) do not full fill these requirements. Currently available solutions are optimized regarding the power efficiency in maximum load scenarios. However, the maximum load scenario is a rather rare case in typical network deployments of operators. That means that most of the time, the network infrastructure is operated with sub-optimal EE thus leading to a significant waste of energy resources. Therefore, power efficiency of base stations components has significant further improvement potential. Radical new approaches should be envisaged to realize fundamental efficiency improvement. To this end, new configurations of PA should be investigated. Actually, PA should be able to tune the output power for different operating conditions in real time. For instance, power variations at the output of the PA cause impedance variations resulting in antenna mismatch and thereby in power wastage of the overall system. In our vision, joint optimization of PA and antenna networks should be introduced to achieve reconfigurable/tuneable interstate matching networks. Moreover, transceivers should support the scaling of the consumed energy in order to enable the adaptation of energy consumption to actual performance requirements. Such energy scalability [28] should be present in all components (analogue RF front-end, baseband) and in all devices. Furthermore, transceivers should enable dynamic power management in order to keep active only their necessary modules during sleep modes. Indeed, at component level, a key investigation axe is the design of traffic load adaptive base station components, with harmonized solutions for EE control on system level and energy efficient base station components unlocking solutions for EE operation of cellular networks.

5.2 Design of Link Interfaces and Application of Innovative Radio Transmission Techniques

State of the art radio link designs (such as beam forming, MIMO, adaptive antennas arrays, coordinated multi-point transmission and, advanced retransmission schemes) have been designed to achieve high peak data rates, good coverage and low latencies. As consequence, mobile terminals require to operate periodic control messages and reference signals from base stations. This limits the possibilities for base station sleep intervals in low load scenarios. Consequently, a significant amount of energy today is wasted by not having appropriate base station sleep mechanisms in place. Some researchers have started proposing solutions on isolated aspects and mainly non-cellular systems [29–32]. A similar embracing analysis of the EE of radio link technologies for cellular is lacking, despite cellular systems represent the major energy consumers due their universal proliferation. In particular, the aspects of the radio link operation in low load scenarios have not been addressed so far. In order to progress the state of the art on energy efficient radio link technologies, several investigation axes should be explored. First, innovation should achieve joint optimization of energy efficient radio link operation strategies for all load regimes. Second, the specific case of low load and short range coverage such as in femto cells deployment should be extensively investigated.

6 Green Adaptive Scheduling for OFDMA Based Systems

Traditional packet scheduling algorithms are designed to increase the maximum system capacity, subject to QoS constraints and fairness. Nevertheless, most of the time, wireless systems are only moderately loaded. With state of the art priority scheduling algorithms, the exploitation of available time and frequency resources is typically not optimized from an energy perspective. The goal of our investigation is to design an energy efficient scheduling algorithms which meets the QoS constraint of an heterogeneous population of UEs. In system load scenarios that are not saturated, the exploitation of available time and frequency resources is typically not optimized from an energy perspective. Based on these observations, we propose to split the resource allocation process into four steps:

Step 1: The Rushing Entity Classifier (REC) classifies *entities* (packets of UEs) waiting to be scheduled as *rushing* or *non-rushing*.

With real-time (RT) traffic, packets are classified as *rushing* if $Th_{rush} \cdot TTL + \eta \geq R_{TTL}$, where Th_{rush} is a threshold on the QoS deadline that depends on the traffic type, TTL is the time to live of the packet, η is a constant which takes into account both retransmission interval and maximum allowed number of retransmissions, and R_{TTL} is the remaining TTL.

With non-real-time (NRT) traffic, flows and not packets are classified by the REC. Therefore, the ith user equipment (UE_i) is classified as *rushing* if one of its NRT flows has been under-served during TW_i. More precisely, every Transmission Time Interval (TTI) the REC checks for each UE_i if $(TW_i - t_{now,i}) \leq (QoS_i - tx_{data,i})/R_{min}$, where $t_{now,i}$ is the elapsed time since the beginning of TW_i, QoS_i the QoS requirements of the UE class of traffic, $tx_{data,i}$ the total data transmitted by user i during $(TW_i - t_{now,i})$ and R_{min} the minimum active transmission rate of the system (always larger than zero). Note that Th_{rush}, η and TW_i are scheduler design parameters.

Step 2: Resources (chunks) are allocated to *rushing entities* with an EDF-like scheduler which allocates *best* chunk(s) to entities with higher deadline priority. Deadline priority metrics differ between RT and NRT traffics: while with RT traffic deadline priority depends on R_{TTL}, with NRT traffic it depends on the under served data rate experienced by UE_i during TW_i.

Step 3: All unscheduled resources (chunks) are assigned to users which maximize spectral efficiency, regardless to any QoS constraints. Priority is granted indeed to UEs selected by standard MCI 'matrix-based chunk allocation' described in [33]. Actually, the matrix contains the metrics $\lambda_{k,n}(i)$ of all possible user-chunk pairs (only for previously unscheduled chunks), where at time i, UE k has a metric for chunk n which is given by $\lambda_{k,n}(i) = R_{k,n}(i)$. $R_{k,n}(i)$ is the instantaneous supportable rate for UE k at chunk n, depending on each UE's Channel Quality Indicator (CQI). At each time i, given an unscheduled chunk n, the scheduled UE is $U_n(i) = \text{argmax}_k \lambda_{k,n}(i)$. Indeed, the Adaptive Modulation and Coding

(AMC) algorithm suggests for each UE_k assigned to chunk n the most spectral efficient Modulation and Coding Scheme (MCS), $MCS^*_{k,n}$, which meets PER_{target}.

Step 4: We trade off UE's throughput for downlink power by limiting the power budget on each *non-rushing* allocated chunk. Therefore, in order to meet the QoS constraints while lowering the transmission power budget, a lower MCS order has to be selected for transmission. Two solutions are possible. Either power is lowered to a given *low power* level, or a target low power MCS is fixed: $MCS_{low-power}$. Such $MCS_{low-power}$ can be for instance suggested by the admission control based on the actual traffic load of the cell. Then, for each UE_k assigned to chunk n, the scheduler scales down $MCS^*_{k,n}$ to low power MCS ($MCS_{low-power}$) only if $MCS^*_{k,n} > MCS_{low-power}$. Downlink transmission power is indeed reduced for chunks allocated to *non-urgent* UEs.

6.1 Simulation Results

In this section we assess the effectiveness of our proposed GAS algorithm (see Section 6) comparing it with two scheduling algorithms often investigated in the literature: MCI, and EDF. Schedulers are compared in different traffic load scenarios in terms of the following power efficiency metric Γ_i:

$$\Gamma_i = \frac{\sum_{j=1}^{K_i^{UE}} N_{i,j}^{RB} \cdot P_{i,j}}{\sum_{j=1}^{K_i^{UE}} \rho_{i,j}} \quad (1)$$

where at TTI i, K_i^{UE}, $N_{i,j}^{RB}$, $P_{i,j}$ and $\rho_{i,j}$ are respectively the number of active UEs in the cell, the number of chunk allocated to user j, the downlink power for transmission on each chunk and the throughput of user j.

Schedulers are compared in three traffic scenarios:

- *Scenario A* (*single real-time traffic*): unique VoIP [34] or NRTV [35] traffic type in the cell for all UEs.
- *Scenario B* (*mixed real-time traffic*): coexistence of VoIP and NRTV traffic in the same cell.
- *Scenario C* (*heterogeneous mixed traffic*): coexistence of VoIP and HTTP [35] traffic in the same cell.

Simulation results are given for the system and traffic models summarized in Table 1. Note that we simulate the traffic of the central cell, while others

Table 1 Main system model parameters.

Network	
Parameter	Value
Carrier frequency	2.0 GHz
Bandwidth	10 MHz
Inter-site distance	500 m
Minimum distance	35 m
TTI duration	1 ms
Cell layout	Hexagonal grid, 19 three-sectored cells
Link to System interface	EESM
Traffic model	VoIP, NRTV
Nb of antennas (Tx, Rx)	(1,1)
Access Technique	OFDMA
Total Number of Sub-carriers	600
Nb of Sub-carriers per PRB	12
Total Nb of Chunks (PRB)	50
Propagation Channel	
Fast fading	Typical urban 6-tap model, 3 km/h
Interference	White
UE	
Channel estimation	ideal
Turbo decoder	max Log-MAP (8 iterations)
Dynamic Resource Allocation	
Nb of MCS	12 (from QPSK 1/3 to 64-QAM 3/4)
AMC PER_{target}	10 %
CQI report	Each TTI, with 2 ms delay
Packet Scheduling	EDF, MCI, and GAS
Sub-carriers Allocation Strategy	Chunk based allocation

BSs are used for down-link interference generation only. Results are averaged over 100 independent dynamic runs, where at the beginning of each run UEs are randomly uniformly located in the central cell. Each run simulates 50 seconds of network activity and at each TTI channel realizations are updated. We check if QoS is met based on the metrics defined in [34, 36]. Real-time VoIP and NRTV UEs are satisfied if more than 95% of the UEs have a residual Packet Error Rate (PER) below 2% and their TTL is respectively 50 and 100 ms.

In Figure 4 we show our simulation results for scenario A with single VoIP traffic and we consider a number of active UEs in the cell that ranges between 20 to 540. Three scheduling algorithms are investigated under this scenario: EDF (dashed curve), MCI (dotted curve) and GAS (solid curve). Note that 540 satisfied VoIP UEs is the limit with any of the investigated schedulers. The highest system load (540 satisfied active UEs) is achieved with EDF and GAS only. MCI can satisfy up to 450 coexistent active UEs. Such maximum load gap between EDF and MCI is not surprising. Actually, since MCI allocates resources to UEs with better momentary link quality regardless of the user time QoS constraints, with the increasing number of real-time flows, many users may face momentary service starvation and consequently, exceed the maximum delivery delay (50 msec). This is not the case with EDF since it allocated best chunk(s) to entities with higher QoS deadline priority. From an EE point of view, we observe that EDF performs slightly better than MCI. Actually, UEs served with MCI experience an higher average transmission spectral efficiency. This causes an higher energy loss caused by the transmission of padding bits added when a UE has not enough data to entirely fill the allocated chunk(s). Our simulation results show how GAS algorithm permits to notably improve the energy effectiveness. We observe that in low load scenarios (20 active UEs), the Γ reduction is up to 32.5%, and it decreases as expected for higher number of active UEs, having up to 31, 29, 26 and 21% of energy cost reduction with respectively 50, 100, 200 and 350 (up to 550) active UEs.

In Figures 5 and 6 we show our simulation results for scenario B and C, respectively when 75 NRTV (or 100 HTTP in Figure 6) UEs are active and the number of active VoIP UEs ranges between 0 and 250 (or between 0 and 450 in Figure 6). This is a more challenging scenario since the system is heavily loaded and heterogeneous QoS constraints have to be taken into account in the scheduling rule. As in Figure 4, we compare Γ performance of EDF (dashed curve), MCI (dotted curve) and GAS (solid curve). In case of Figure 5 while MCI cannot satisfy any additional VoIP UEs when 75 NRTV UEs are present in the system (therefore only a dot is represented in the figure), EDF and GAS can satisfy respectively up to 220 and 250 additional VoIP UEs. Simulation results show that comparing EDF and GAS, GAS improves Γ performance respectively of 18, 16, and 10.5%, with respectively up to 50, 150 and 200 VoIP active UEs. With 250 VoIP UE only GAS can satisfy all active UEs since it is more flexible and effective in mixed traffic scenarios.

Finally, in Figure 6 we show our simulation results for scenario C, when 100 HTTP UEs are active and the number of active VoIP UEs ranges between 0 and 450. Again, heterogeneous QoS constraints have to be taken into account in the scheduling rule. With the same trend of results, in Figure 6, while EDF and MCI perform slightly the same (MCI can serve more VoIP UEs), best performance is obtained with GAS. We observe with GAS scheduling a Γ performance improvements of respectively 30 and 15% with 100, and

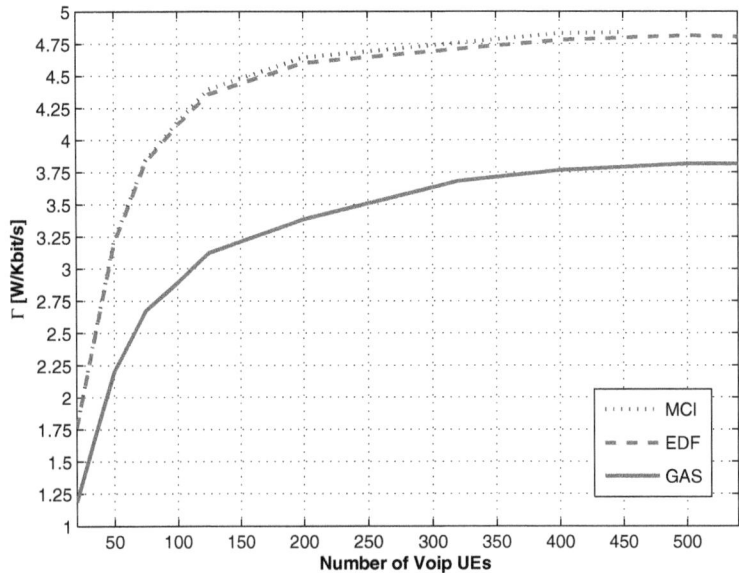

Figure 4 Scenario A (single traffic): Γ performance evaluation with EDF, MCI and GAS schedulers varying the load of VoIP UEs.

200 VoIP active UEs. Furthermore, in scenario C GAS is able to serve up to 450 VoIP, while EDF and MCI can satisfy respectively up to 150 and 250 VoIP UEs. Note that GAS presents some steps in its performance since increasing the number of active users, a smaller room for modulation scaling of non-urgent UE is experienced.

7 Ghost Femtocells for Energy Efficient Heterogeneous Cellular Network

In our view, femtocells should be invisible in terms of interference generated to neighbour cellular users. Nevertheless, femtocells deployment presents a very challenging issue: while HeNBs power consumption and *interference range* should be *small*, the *coverage range* at which UEs can meet their QoS constraints should be *large*. Based on this observation, we propose a novel RRM algorithm designed to strongly lower the HeNBs downlink transmission power. We come out with the design of an RRM algorithm, the *Ghost Femtocells* (RRM_{Ghost}), which trades off transmission energy for

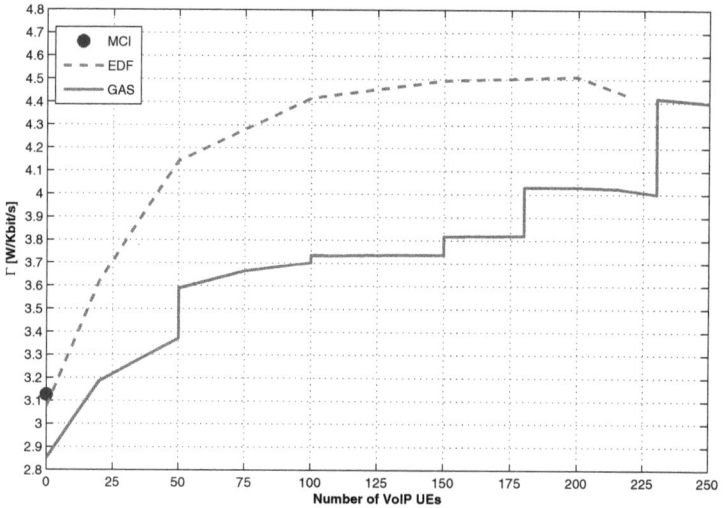

Figure 5 Scenario B (mixed traffic): Γ performance evaluation with EDF, MCI and GAS schedulers with 75 NRTV UEs varying the load of VoIP UEs.

frequency resources. The detailed description of the proposed algorithm is as follows:

Step 1 (Classification of Interferers): H-UEs overhear the Broadcast CHannel (BCH) and estimate which neighbour HeNBs are currently *strong* interferers. A interferer is *strong*, if its sensed power level is larger than a predefined threshold.

Step 2 (Feedback to HeNB): H-UE feedbacks to the HeNB its QoS constraints, the momentary Channel State Indicator (CSI) measurements, and the cell-IDs of the HeNBs perceived as *strong* interferers.

Step 3 (Feedback to Control Unit): Each HeNB within the femtocell network (i.e. the group of femtocells placed in a block of apartments) reports this information to the Control Unit (CU). For each user i in the network, the CU stores the set of its neighbours V_i. The elements of this set are the users that are served by the HeNB of user i ($HeNB^i$) and the users that are served by the HeNBs that are indicated as strong interferers by $HeNB^i$.

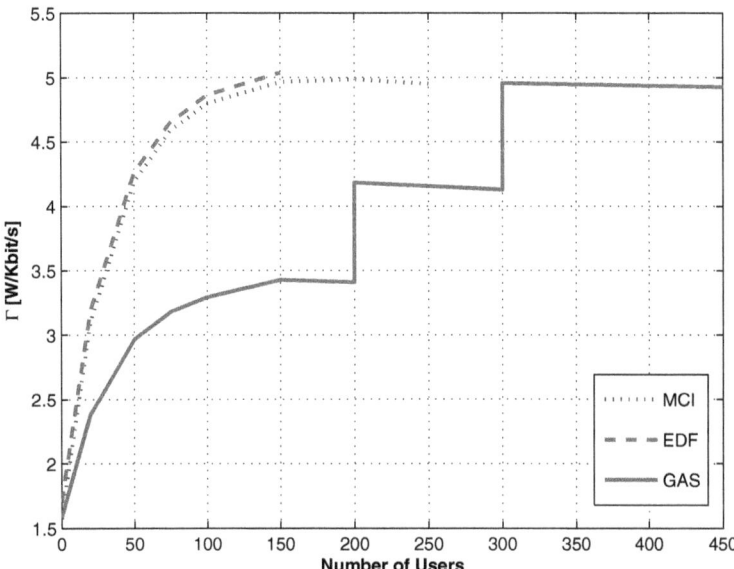

Figure 6 Scenarios C (mixed traffic): Γ performance evaluation with EDF, MCI and GAS schedulers with 100 HTTP UEs varying the load of VoIP UEs.

Step 4 (Computing Scheduling Matrices): According to the CSI measurements and the selected scheduler algorithm, the CU computes scheduling metric λ_i^j for every user i on every RB j.

In this paper, the RRM_{Ghost} implements a proportional fair based scheduler, that is

$$\lambda_i^j = SINR_i^j \bigg/ \sum_{k=1}^{K} SINR_i^k, \qquad (2)$$

where $SINR_i^j$ represents the instantaneous channel condition of the RB j observed at user i and $\sum_{k=1}^{K} SINR_i^k$ is the sum of SINR of K RBs that have been already allotted to user i. RRM_{Ghost} uses this metric to build the scheduling matrices M^{Tx} and M^{Rep} of dimension $[\sum_{k=1}^{N_f} N_k \times N_{RB}]$, where N_f is the number of active HeNBs in the network, N_k is the number of users served by the femtocell k, and N_{RB} is the number of available RBs. Based on M^{Tx}, the scheduler allocates to each user the minimum number of RBs that meets QoS and power constraints. Then, the proposed scheduler sorts matrix

M^{Rep} to allocate to the served users additional available RBs.

Step 5 (Scheduling): The CU selects the minimum number of RBs that meets QoS and power constraints for each user to be served. It schedules in three steps:

Step 5-a: The controller selects the best user-available RB pair (i, j) with the best metric in M^{Tx}.

Step 5-b: For each user-available RB pair (i, j), the algorithm checks the set of RBs allotted to user i ($\widehat{RB}_i$) and selects the highest possible Modulation and Coding Scheme ($\widehat{MCS}_i$), accordingly. The overall available power at user i served by the HeNB k is $\hat{P}_i = P^T/N_k$, where P^T and N_k are the power budget and the number of users of the HeNB k, respectively. The controller equally splits $\hat{P}_i$ in the set $\widehat{RB}_i$.

Step 5-c: Then, the controller estimates the sum of the Mutual Information I given by set $\widehat{RB}_i$ and $\widehat{MCS}_i$.
- When $I = 0$, the selected user-RB pair cannot be served in this scheduling period so the values of the *i-rows* in both M^{Tx} and M^{Rep} are set to zero.
- When $I \geq R_{tg}$, user i is served. The values of the *i-row* in M^{Tx} and $M^{Rep}(i, j)$ are set to zero and the values of the *i-row* in M^{Rep} are updated according to the scheduler rule (see Eq. (2)). Moreover, $M^{Tx}(k, j)$ and $M^{Rep}(k, j)$, where $k \in V_i$, are set to zero.
- If $I < R_{tg}$, the user i is not served yet. The values $M^{Tx}(i, j)$ and $M^{Rep}(i, j)$ are set to zero and the values of the *i-rows* in M^{Rep} and M^{Tx} are updated according to the scheduler rule (see Eq. (2)). Moreover, $M^{Tx}(k, j)$ and $M^{Rep}(k, j)$, where $k \in V_i$, are set to zero.

Step 6 (MCS Scaling): Given the set of RBs ($\widehat{RB}_i$) allocated each served user i, the algorithm finds the MCS^* of the minimum order that meets the QoS target. If MCS^* is different from $\widehat{MCS}_i$, the MCS of user i (MCS_i) is set equal to MCS^*. The goal of this process is twofold. First, it improves the transmission robustness. Second, it reduces the padding thus improving the spectral efficiency.

Step 7 (Spreading):
The CU allocates unused RBs to spread the original message and improve

the transmission robustness. Scheduling is done in three steps:

Step 7-a: The scheduler selects the user-available RB pair (i, j) that has the best metric in M^{Rep}.

Step 7-b: For each user-available RB pair (i, j), the algorithm checks the Mutual Information I given by the entire set of RBs allocated to user i and MCS_i:
- If $I < R_{tg}$, exploit further RBs would cause outage, hence the values of the row corresponding to user I in M^{Rep} are set to zero.
- When $I \geq R_{tg}$, the original message is spread in the additional RB and $M^{Rep}(i, j)$ as well as $M^{Rep}(k, j)$, where $k \in V_i$, are set to zero.

Moreover the values of the *i-row* in M^{Rep} are updated according to the scheduler rule.

Step 7-c: The scheduler process terminates when no more user-RB pairs are available.

Step 8 (Power Scaling): The algorithm estimates the SINR perceived at each served user and reduces the allocated transmission power to meet the SINR threshold given by the target PER and the selected MCS.

Step 9 (Message Reception): Finally, each user collects the information received in each of its allotted RBs and combines these RBs using the chase combining scheme [37].

7.1 Deployment of Femtocells and System Model

In our analisys, we assume that femtocells are deployed according to the 3GPP grid urban deployment model [38] (see Figure 7). This model represents a single floor building with 10 m × 10 m apartments placed next to each other in a 5 × 5 grid. The block of apartments belongs to the same region of a macrocell. Moreover, we assume that six additional M-BSs surround the central macrocell generating additive interference for both macro and femto users. Each HeNB can simultaneously serve up to four users. In order to consider a realistic case in which some apartments do not have femtocells, we use a system parameter ρ_d called a deployment ratio that indicates the percentage of apartments with a femtocell. Furthermore, the 3GPP model

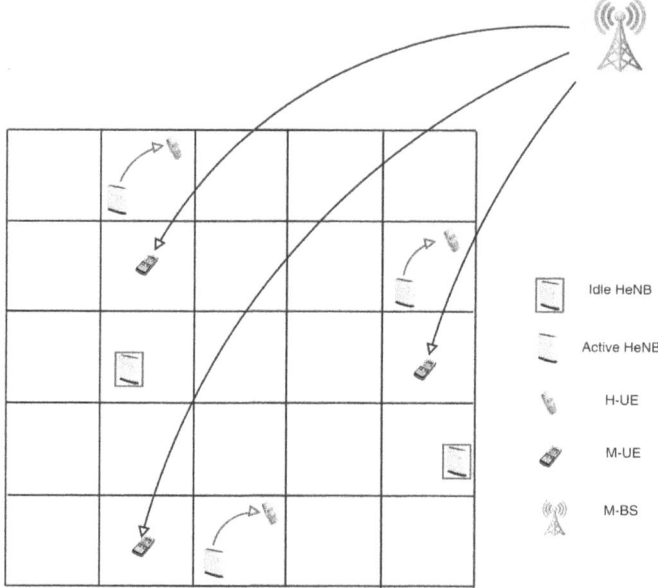

Figure 7 3GPP femtocell grid urban deployment model.

includes ρ_a, another parameter called an activation ratio defined as the percentage of active femtocells. If a femtocell is active, it will transmit with a certain power over data channels. Otherwise, it will transmit over the control channel.

We use two different models to capture the signal propagation effect based on the 3GPP specifications [38]:

1. Transmissions from femto and macro users to M-BSs:

$$PL(dB) = 15.3 + 37.6 \log_{10} d + L_{ow},$$

where d is the distance between the transmitter and the receiver (in meters) and L_{ow} is the penetration losses of an outside wall equal to 20 dB.

2. Transmissions from femto and macro users to HeNBs:

$$PL(dB) = 38.46 + 20 \log_{10} d + 0.7 d_{2D,indoor} \\ + 18.3 n^{((n+2)/(n+1)-0.46)} + q \cdot L_{iw},$$

where d is the distance between the transmitter and the receiver (in meters), $d_{2D,indoor}$ is the two dimensional separation between the trans-

Table 2 Main system model parameters.

Parameter	Value
Carrier frequency	2.0 GHz
Carrier bandwidth	10.0 MHz
Total number of Resource Blocks	50
Inter-site distance	500 m
Access Technique	OFDMA
M-BS Tx power	16 dBW
M-BS antenna gain after cable loss	13 dB
HeNB maximum Tx power	−17 dBW
HeNB antenna gain after cable loss	0 dB
Shadowing distribution	Log-normal
Shadowing standard deviation in the interfering links	8 dB
Shadowing standard deviation in the femto/macro user useful link	4/8 dB
Autocorrelation distance of shadowing	50 m
Fast fading distribution	Rayleigh
Thermal noise density	$N_0 = -174$ dBm/Hz

mitter and the receiver (in meters), n is the number of penetrated floors, q is the number of walls that separate the user apartments and the transmitting HeNB apartment, L_{iw} is the penetration loss of walls within the grid of apartments equal to 5 dB. The third term in the above expression represents the penetration loss due to walls inside an apartment. This attenuation is modeled as a log-linear value equal to 0.7 dB/m. The fourth term represents the penetration loss within different floors. In the single floor building scenario considered, we have $d = d_{2D,indoor}$ and $n = 0$.

Table 2 shows key model parameters including shadowing, fast fading, the macrocell antenna gain, and the transmission power.

7.2 Simulation Results

In this section, we assess the effectiveness of the proposed RRM_{Ghost} by comparing its performance with a reference algorithm ($RRM_{classic}$). The main differences between these schemes are:

1. In $RRM_{classic}$, there is no coordination within the femtocell network. Hence, HeNBs are not aware of the presence and allocation strategy of neighbour HeNBs.
2. $RRM_{classic}$ aims to maximize the spectral efficiency of femtocells while minimizing the probability that users that belong to different cells access

to same RBs. Thus, the $RRM_{classic}$ attempts to limit the number of RBs allotted to each H-UE.
3. $RRM_{classic}$ algorithm does not implement MCS and Power scaling (Steps 6 and 8 in RRM_{Ghost} algorithm).

We present simulation results for the system model and its parameters presented in Section 7.1. RRM algorithms are compared in terms of the efficiency metric defined in Section 6.1, Eq. (1).

The results are averaged over 100 independent runs. For each run 10^3 independent TTIs are simulated and at each TTI channel fading instances are updated. At the beginning of each run, we randomly place two blocks of apartments where HeNBs, H-UEs, and M-UEs are randomly deployed. Note that in the presented simulations, we consider that all deployed HeNBs are active ($\rho_a = 1$) with four H-UEs per HeNB. Moreover, indoor M-UEs are randomly distributed in the apartments where HeNBs are not deployed.

In our simulations, solid and dashed lines, respectively, correspond to the performance of $RRM_{classic}$ and RRM_{Ghost} schemes.

Figure 8 shows the H-UEs performance as Γ versus the power budget P^T at each femtocell. We have set ρ_d equal to 0.3 and considered four different traffic scenarios:

Scenario Traf.1: H-UE throughput target $T_{tg} = 300$ kbit/s, square marked curves.

Scenario Traf.2: H-UE throughput target $T_{tg} = 600$ kbit/s, circle marked curves.

Scenario Traf.3: H-UE throughput target $T_{tg} = 1$ Mbit/s, star marked curves.

Scenario Traf.4: H-UE throughput target $T_{tg} = 2$ Mbit/s, diamond marked curves.

From Figure 8 we can observe that the proposed RRM_{Ghost} improve the performance of H-UEs compared to $RRM_{classic}$ in all considered traffic scenarios. For instance, considering a HeNB power budget equal to 10 mW, the proposed RRM_{Ghost} gain up to 94.5% with respect to $RRM_{classic}$ in Scenario Traf.1, up to 90% in Scenario Traf.2, up to 85% in Scenario Traf.3, and up to 75.4% in Scenario Traf.4. Moreover, results outline how the gain increases in lightly loaded scenarios (Scenario Traf.1 and Scenario Traf.2) where our

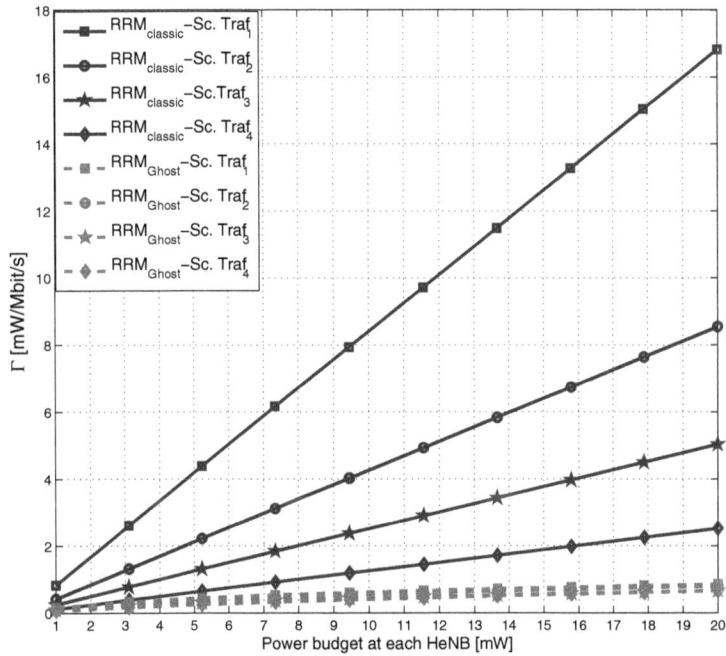

Figure 8 Average Γ measured at H-UEs as function of power budget at each HeNB in different traffic scenarios.

algorithms permit to strongly reduce the transmission power by improving the femtocell energy efficiency. In this cases, our schemes takes advantage of the lower throughput targets to decrease the downlink transmission power and spreading over RBs.

Figures 9 and 10 show the improvement in M-UE performance under RRM_{Ghost}. In the co-channel femtocell deployment, indoor M-UE performance is limited by *femto-to-macro interference*. Some recent research introduced cooperation within M-BSs and HeNBs in order to coordinate the access to the radio medium and avoid the *cross-tier interference* [39, 40]. However, following the 3GPP Release 10 [41], we do not implement this coordination in our system. Hence, the M-BS scheduler is not aware of the RBs exploited by the interfering HeNBs. When the M-BS assigns an RB to an indoor user which is used by a neighbour HeNB, this M-UE can be exposed to a high level of interference. We aim to evaluate the effect of this interference on M-UE when femtocells use the reference $RRM_{classic}$ and the proposed scheme.

In order to compare these algorithms, we have set the M-UE throughput target equal to 300 kbit/s, the H-UE throughput target equal to 600 kbit/s and considered three different femtocell deployment scenarios:

Scenario δ_L: low density – $\rho_d = 0.3$, circle-marked curves.

Scenario δ_M: medium density – $\rho_d = 0.5$, triangle-marked curves.

Scenario δ_H: high density – $\rho_d = 0.8$, plus-marked curves.

In Figure 9 we show the indoor M-UE performance as Γ versus the power budget P^T at each femtocell. Results indicate that RRM_{Ghost} permit to limit the impact of the *femto-to-macro interference* in all scenarios. For instance, considering P^T equal to 20 mW (20 mW is often indicated by researchers and industries as the current downlink power target for HeNBs), the proposed RRM_{Ghost} gains up to 10.6, 16.2, and 21.5% with respect to $RRM_{classic}$ in Scenarios δ_L, δ_M, and δ_H. In low/medium density scenarios, the probability that neighbour HeNBs allocate the same RBs at the same time is fairly small under $RRM_{classic}$. Hence, RRM_{Ghost} improvements mainly come from steps 6, 7, and 8 of the proposed schemes (MCS scaling, spreading, and power scaling) that permit to reduce the HeNB downlink power transmission in each RB. However, in Scenario δ_H, several neighbour HeNBs may access same RBs in the same TTI resulting in high peak of interference on some RBs. In RRM_{Ghost}, the CU coordinate the access of neighbouring femtocells to further limit the overall interference perceived at end-users of both macro and femto cells.

Figure 10 shows the M-UE performance as a function of distance between the end-user and the M-BS for different femtocell deployment scenarios and for P^T equal to 20 mW. M-UEs that are located far away from the M-BS perceive low average SINR due to path loss and interference generated by surrounding M-BSs. Hence, it is fundamental to limit the *femto-to-macro interference* to satisfy these macro users' QoS constraints. Simulation results show that the proposed algorithm strongly enhance the performance of M-UEs placed at the border of the cell. Furthermore, the observed gain increases with higher density of femtocells and larger distance between the M-UE and its M-BS.

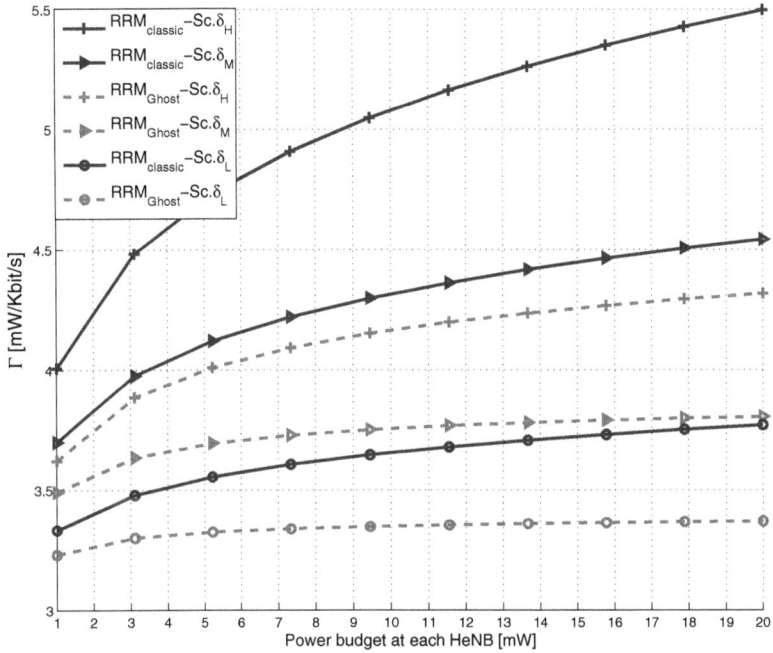

Figure 9 Average Γ measured at M-UEs as a function of the power budget at each HeNB in different traffic scenarios.

In Scenario δ_L, under $RRM_{classic}$, the M-BS needs 5.8 mW/Kbit/s to serve a M-UE placed at 225 meters far away, while only 4.6 mW/Kbit/s are necessary with the proposed RRM_{Ghost}. In Scenario δ_M, under $RRM_{classic}$, the M-BS needs 7.7 mW/Kbit/s to serve a M-UE placed at 225 meters far away, while only 5.5 mW/Kbit/s are necessary with the proposed RRM_{Ghost}. In Scenario δ_H, under $RRM_{classic}$, the M-BS needs 10.9 mW/Kbit/s to serve a M-UE placed at 225 meters far away, while only 7.1 mW/Kbit/s are necessary with the proposed RRM_{Ghost}.

This improvements result in higher macrocell coverage and capacity. The maximum number of indoor M-UEs placed at the border of the cell that can be served simultaneously is

$$N_{M-UE} = \frac{P^T_{M-BS}}{\Gamma \cdot T^{M-UE}_{tg}}, \qquad (3)$$

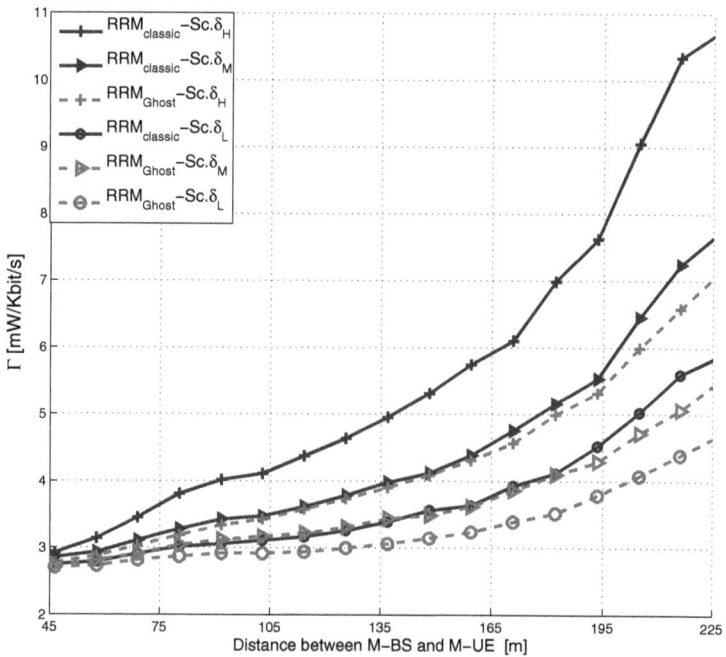

Figure 10 Average Γ measured at M-UEs as a function of the distance between the end-user and the M-BS.

Table 3 Maximum number of indoor M-UEs placed at the border of the macrocell that can be contemporary served by an M-BS.

ρ_d \ RRM algorithm	$RRM_{classic}$	RRM_{Ghost}
0.3	19	21
0.5	13	18
0.8	9	14

where P_{M-BS}^T and T_{tg}^{M-UE} are respectively the M-BS downlink available power and the M-UE throughput target. Table 3 shows the values of N_{M-UE} under the discussed algorithms and in different deployment scenarios (P_{M-BS}^T and T_{tg}^{M-UE} are set equal to 16 dBW and 300 Kbit/s, respectively).

8 Conclusion

Communication Energy Efficiency (EE) presents indeed an alarming bottleneck in the telecommunication growth paradigm. In this paper, we first detailed the holistic energy efficient optimization methodology for broadband cellular networks proposed by the EARTH project. We discussed the new research activities that are currently investigated in the vivid area of *Green Communications*. The topics of energy efficiency metrics, network deployment strategies, network management, and energy efficient radio resource management have been introduced. Then, we investigated downlink power effectiveness of two reference scheduling algorithms MCI and EDF, in a systems that implements a realistic OFDMA air interface where no full-queued traffic and unlimited number of control channel per TTI is assumed. We underlined that while EDF does not profit from multi-user diversity, MCI targets at maximizing the cell throughput regardless of the user's QoS constraints even in low load scenarios for which spectral efficiency can be traded with energy efficiency. Then, we came up with the definition of a novel scheduler, the GAS algorithm. We evaluated the effectiveness of the proposed GAS algorithm comparing it with the above reference schedulers. Our simulations substantiated how GAS is a highly flexible and effective scheduler for a variety of traffic scenarios and it drives to notable energy cost reductions while meeting the Quality of Service (QoS) of UEs active admitted in the system. In the third part of the paper, we presented the RRM_{Ghost}, a novel radio resource management scheme that efficiently uses the available wireless spectrum in a two-tier network. In this scenario, a large population of heterogeneous interferers shares the overall cellular bandwidth. The proposed algorithm limits the undesired effects of interference and allows a drastic reduction of the downlink energy per information bit required to satisfy the QoS constraints of both macro and femto users.

We expect that both the holistic system methodology and the proposed energy efficient resource allocation algorithms presented in this paper will encourage the research community to investigate into innovative solutions in order to achieve a sustainable increase of broadband communication system.

Acknowledgements

The authors gratefully acknowledge the contribution of the EARTH consortium to the technical definition of the project. The work leading to this paper has received funding from the European Community's Seventh Framework Programme [FP7/2007-2013] under Grant No. 247733 project EARTH.

References

[1] M. Etoh, T. Ohya, and Y. Nakayama. Energy consumption issues on mobile network systems. In *Proceedings of the International Symposium on Applications and the Internet, SAINT 2008*, pp. 365–368, July–August 2008.
[2] S. Fletcher. Green radio-sustainable wireless networks. *Mobile VCE Core*, 5, February 2009. Online: http://www.mobilevce.com/dloadspubl/mtg284Item1503.ppt.
[3] Mobile VCE, http://www.mobilevce.com/.
[4] GreenTouch, http://www.greentouch.org/.
[5] M. Gruber, O. Blume, D. Ferling, D. Zeller, M.A. Imran, and E. Calvanese Strinati. EARTH – Energy Aware Radio and Network Technologies. In *Proceedings of IEEE 20th International Symposium on Personal, Indoor and Mobile Radio Communications (PIMRC 2009)*, pp. 1–5, 2009.
[6] Energy Aware Radio and neTworking tecHnologies European Community's Seventh Framework Programme, https://www.ict-earth.eu/.
[7] E. Calvanese Strinati and P. Greco. Green resource allocation for ofdma wireless cellular networks. In *Proceedings of IEEE International Symposium on Personal, Indoor and Mobile Radio Communications*, Instanbul, Turkey, September 2010.
[8] V. Chandrasekhar, J. Andrews, and A. Gatherer. Femtocell networks: A survey. *IEEE Communications Magazine*, 46(9):59–67, September 2008.
[9] Airvana Inc. How femtocells change the economics of mobile service delivery, http://www.airvana.com/.
[10] M. Heath et al. Picocells and femtocells: Will indoor base stations transform the telecoms industry? http://research.analysys.com, 2007.
[11] S. Carlaw. IPR and the potential effect on femtocell markets. FemtoCells Europe, ABIresearch, 2008.
[12] S. Carlaw and C. Wheelock. Femtocell market challenges and opportunities. ABI Research, Research Report, Vol. 23, 2007.
[13] B. Badic, T. O'Farrrell, P. Loskot, and J. He. Energy efficient radio access architectures for green radio: Large versus small cell size deployment. In *Proceedings of IEEE 70th Vehicular Technology Conference Fall (VTC 2009-Fall)*, pp. 1–5, 2009.
[14] Y. Chen, S. Zhang, and S. Xu. Characterizing energy efficiency and deployment efficiency relations for green architecture design. In *Proceedings of IEEE International Conference on Communications Workshops (ICC 2010)*, pp. 1–5, 2010.
[15] I. Ashraf, L.T.W. Ho, and H. Claussen. Improving energy efficiency of femtocell base stations via user activity detection. In *Proceedings of IEEE Wireless Communications and Networking Conference (WCNC 2010)*, pp. 1–5, 2010.
[16] C.H.M. de Lima, M. Bennis, K. Ghaboosi, and M. Latvaaho. Interference management for self-organized femtocells towards green networks. In *Proceedings of IEEE 21st International Symposium on Personal, Indoor and Mobile Radio Communications Workshops (PIMRC2010)*, pp. 352–356, 2010.

[17] A. De Domenico and E. Calvanese Strinati. A radio resource management scheduling algorithm for selforganizing femtocells. In *Proceedings of IEEE 21st International Symposium on Personal, Indoor and Mobile Radio Communications Workshops (PIMRC2010)*, pp. 191–196, 2010.

[18] Emilio Calvanese Strinati, Antonio De Domenico, and Andrzej Duda. Ghost femtocells: A novel radio resource management scheme for OFDMA based networks. In *Proceedings of IEEE Wireless Communications and Networking Conference (WCNC 2011)*, Cancun, Mexico, March 2011.

[19] B. Gorjni. ETSI work programme on energy saving. In *Proceedings of 29th International Telecommunications Energy Conference (INTELEC2007)*, pp. 174–181.

[20] G. Auer, I. Gódor, L. Hvizi, M.A. Imran, J. Malmodin, P. Fazekas, G. Biczók, H. Holtkamp, D. Zeller, O. Blume, and R. Tafazolli. Enablers for energy efficient wireless networks. In *Proceedings of IEEE 72nd Vehicular Technology Conference Fall (VTC 2010-Fall)*, pp. 1–5, 2010.

[21] Ambient Networks, EU FP6 IP, phase 1 2004–2005, phase 2 2006–2007, http://www.ambientnetworks.org/.

[22] R.G. Gallager. Power limited channels: Coding, multiaccess, and spread spectrum. In *Proceedings of 1988 Conference Information Sciences and Systems*, March 1988.

[23] S. Verdu. On channel capacity per unit cost. *IEEE Transactions on Information Theory*, 36(5):1019–1030, 1990.

[24] C. Schurgers, O. Aberthorne, and M.B. Srivastava. Modulation scaling for energy aware communication systems. In *Proceedings of International Symposium on Low Power Electronics and Design*, 2001.

[25] F.M. Chiussi and V. Sivaraman. Achieving high utilization in guaranteed services networks using early deadline-first scheduling. In *Proceedings of Sixth International Workshop on Quality of Service (IWQoS1998)*, pp. 209–217, May 1998.

[26] A. Pokhariyal, T.E. Kolding, and P.E. Mogensen. Performance of downlink frequency domain packet scheduling for the UTRAN long term evolution. In *Proceedings of IEEE 17th International Symposium on Personal, Indoor and Mobile Radio Communications (PIMRC2006)*, pp. 1–5, 2006.

[27] F. Meshkati, H.V. Poor, S.C. Schwartz, and N.B. Mandayam, An energy-efficient approach to power control and receiver design in wireless data networks. *IEEE Transactions on Communications*, 53(11):1885–1894, 2005.

[28] L. Benini, A. Bogliolo, and G. De Micheli. A survey of design techniques for system-level dynamic power management. *IEEE Transactions on Very Large Scale Integration (VLSI) Systems*, 8(3):299–316, 2000.

[29] Jui-Hung Yeh, Jyh-Cheng Chen, and Chi-Chen Lee, Comparative analysis of energy-saving techniques in 3GPP and 3GPP2 systems. *IEEE Transactions on Vehicular Technology*, 58(1):432–448, 2009.

[30] Shuguang Cui, A.J. Goldsmith, and A. Bahai. Energy efficiency of MIMO and cooperative MIMO techniques in sensor networks. *IEEE Journal on Selected Areas in Communications*, 22(6):1089–1098, 2004.

[31] A.J. Goldsmith and S.B. Wicker. Design challenges for energy-costrained ad hoc wireless networks. *IEEE Wireless Communications*, 9(4), 2002.

[32] G. Miao, N. Himayat, Y.G. Li, and A. Swami, Crosslayer optimization for energy-efficient wireless communications: A survey. *Wireless Communications and Mobile Computing*, 9(4):529–542, 2009.
[33] V. Ramachandran, V. Kamble, and S. Kalyanasundaram. Frequency selective OFDMA scheduler with limited feedback. In *Proceedings of IEEE Wireless Communications and Networking Conference*, pp. 1604–1609, April 2008.
[34] 3GPP TSG-RAN1#48. Orange Labs, China Mobile, KPN, NTT DoCoMo, Sprint, T-Mobile, Vodafone, and Telecom Italia, R1-070674, LTE physical layer framework for performance verification, v7.1.0, February 2007.
[35] 3GPP TR 25.892 V6.0.0. Feasibility study for Orthogonal Frequency Division Multiplexing (OFDM) for UTRAN enhancement, June 2004.
[36] 3GPP TSG RAN, 3GPP TR.25814. Physical layer aspects for evolved UTRA (Release 7), v7.1.0, September 2006.
[37] D. Chase. Code combining – A maximum-likelihood decoding approach for combining an arbitrary number of noisy packets. *IEEE Transactions on Communications*, 33(5):385–393, May 1985.
[38] 3GPP TSG-RAN4#51. Alcatel-Lucent, picoChip Designs, and Vodafone, R4-092042, Simulation assumptions and parameters for FDD HENB RF requirements, May 2009.
[39] B. Zubin, S. Andreas, A. Gunther, and H. Harald. Dynamic resource partitioning for downlink femto-to-macro-cell interference avoidance. *EURASIP Journal on Wireless Communications and Networking*, 2010.
[40] M.E. Sahin, I. Guvenc, Moo-Ryong Jeong, and H. Arslan. Handling CCI and ICI in OFDMA femtocell networks through frequency scheduling. *IEEE Transactions on Consumer Electronics*, 55(4):1936–1944, 2009.
[41] 3GPP TSG-RAN1#62, R1-105082. Way forward on eICIC for non-CA based HetNets, August 2010.

Biographies

Emilio Calvanese Strinati obtained his Masters degree in 2001 from the University of Rome La Sapienza and his PhD in Engineering Science in 2005 on "Radio link control for improving the QoS of wireless packet transmission". He started working at Motorola Labs in Paris in 2002. Then in 2006 he joined the Centre for Atomic Energy (CEA) in Grenoble as a research engineer. Since 2004 Emilio Calvanese Strinati has been giving lectures at ENST, the University of Rome La Sapienza, and INPG-Grenoble on physical and MAC layer topics. His main research topics are information theory, advanced coding schemes, cooperative communications, scheduling, resources allocation and green ICT for wireless mobile networks. In 2007 he became a PhD supervisor. Emilio Calvanese Strinati has published around 50 papers in international conference proceedings and book chapters, and is the main inventor or co-inventor of more than 20 patents. Since 2009 he is the Deputy Head of Telecommunication program in CEA.

Antonio De Domenico obtained his Masters degree from the University of Rome La Sapienza, Rome, Italy, in 2008. He is currently working on his PhD degree at the CEA/LETI Labs – MINATEC, Grenoble, France. Furthermore, he is involved in the Inter-Carnot TEROPP and ICT-FP7 BeFEMTO projects that respectively aim to successfully develop opportunistic radio and femtocell technologies in order to enable a cost-efficient provisioning of ubiquitous broadband services.

Laurent Herault was born in Tours, France, in 1964. He received his BS degree in electrical engineering and MS degree in control engineering from the Institute National Polytechnique de Grenoble (INPG) in 1987, and his PhD degree in computer science from INPG in 1991. He won the Best Junior Researcher Award from the University of Grenoble, France, in 1990. He is an 'International Expert' at CEA. From 2004, he is Director of the Telecommunications Program of CEA. Since 2010, he is Head of Wireless & Security Labs in CEA-LETI.

A Reality Check on Home Automation Technologies

Poul Ejnar Rovsing[1], Peter Gorm Larsen[1],
Thomas Skjødeberg Toftegaard[1] and Daniel Lux[2]

[1]*Aarhus School of Engineering, Dalgas Avenue 2, 8000 Aarhus, Denmark;
e-mail: {per, pgl, tst}@iha.dk*
[2]*Seluxit ApS, Kattesundet 24, 1. sal, 9000 Aalborg, Denmark;
e-mail: daniel@seluxit.dk*

Received: 7 February 2011; Accepted: 21 March 2011

Abstract

Buildings account for more than 35% of the energy consumption in Europe and there is a political desire to lower the general energy consumption. Therefore a step towards more sustainable lifestyle could be to use home automation to optimize the energy consumption "automatically". One of the challenges to make this a reality is to make this kind of technology available at an affordable price and easy to get installed. In order for non-technical users to invest in this kind of technology it is important to guard against vendor lock-in. Thus, interoperability between devices from different companies will be essential. In this paper we report about some of the remaining challenges before this can become a reality. In addition we reflect upon the future trends that we foresee in this market.

Keywords: home automation, wireless, protocols, energy saving, interoperability.

1 Introduction

Buildings account for more than 35% of the total energy consumption in Europe. While new buildings are more energy efficient and continue to improve, these enhancements are not enough to reach the European Union's (EU) ambitious goal to improve energy efficiency by 20% before 2020. More than 80% of the European buildings standing in 2020 are already built, so to reach EU's goal the existing buildings have to improve their energy efficiency too. Today, the energy consumption in private households is 22% of the total consumption in Denmark and the costs of private energy consumption have almost doubled since 1990 [28, 35]. Therefore it is relevant to reduce the energy consumption in private households and it has been so since the oil crisis in the 1970s. From the late 1980s on, the process has mainly been motivated by the negative consequences of fossil energy use for the environment, such as global warming.

Energy-saving strategies with focus on private households can be distinguished into the following categories [25]:

- *Technical improvements*, e.g. develop washing machines that use less energy to wash the clothes and develop energy saving light sources to replace incandescent light bulbs.
- *Different use of products*, e.g. wash the clothes at a lower temperature and remember to turn off the light when you leave a room.
- *Shift in consumption*. Indirect energy use can be reduced by consuming less energy-intensive products, by shifting expenditures to goods with a lower energy intensity.

The last two strategies require a behavioural change of the consumer and will often lead to decreased comfort or require an additional effort, but they require no initial investment of the consumer and will often save money. Therefore many studies have focused on social or psychological factors related to energy-saving behaviour while other studies focused on the effects of information and various types of feedback [25]. In general, technical solutions are more acceptable than behavioural changes with most consumers [25]. By use of home automation it is possible to combine the strategies. Home automation can be used as a technical solution to lower the use of the energy consuming products such as automatically turning off the light when no one is in a room, or turning off the heating system (e.g. radiators) or air condition when a window is open in a room. The energy reduction achievable by use of home automation is expected to be in the range of 5 to 10%. It will be possible to combine a home automation system with direct feed-back of energy consumption, and Darby [8] presents a review of savings demonstrated by a total of 38 feedback studies worldwide. She reports that almost all of the projects involving direct feedback produce savings of 5% or more.

In case we wish to lower the energy consumption at a global scale by use of home automation the technology has to be extremely widespread to have any measurable effect and this is not the case today. Although home automation systems have existed for many years they are still not in widespread use.

This paper starts off with elaborating upon the technical challenges with home automation in Section 2. Afterwards an overview of relevant communication protocols commonly used in connection with home automation systems is provided in Section 3. This is followed in Section 4 with a description of the project we have carried out trying to bridge the interoperability gaps between some of these protocols. In Section 5 we present the main result of this paper in the form of the lessons learned by trying to create interoperability in real living labs. This is followed by Section 6 which provides an overview of related work, while our expectation for future tendencies in the home automation area are presented in Section 7. Finally Section 8 provides some concluding remarks.

2 Challenges with Home Automation

There are already a lot of electronic devices in private homes with features which can help to manage and reduce energy consumption and improve comfort in the home. Unfortunately, it is not easy for non-technical people to establish a Home Automation (HA) network working in their own home in the way they would like it to. Today this kind of system is mainly used by technology freaks and wealthy people who can afford to pay a technician to install the system. The challenge really is to enable the people with few financial resources to purchase desirable sensors and actuators for a home automation system from different suppliers and potentially using different communication technologies and different protocols and then establish the interoperability and configure it to meet the desires of the residents (see Figure 1). This means that it is worthwhile to investigate whether it is possible to develop a system that is easy for non-technically minded people to install and configure. And later extend with new devices as the needs arise.

In order to investigate this opportunity a number of challenges must be addressed:

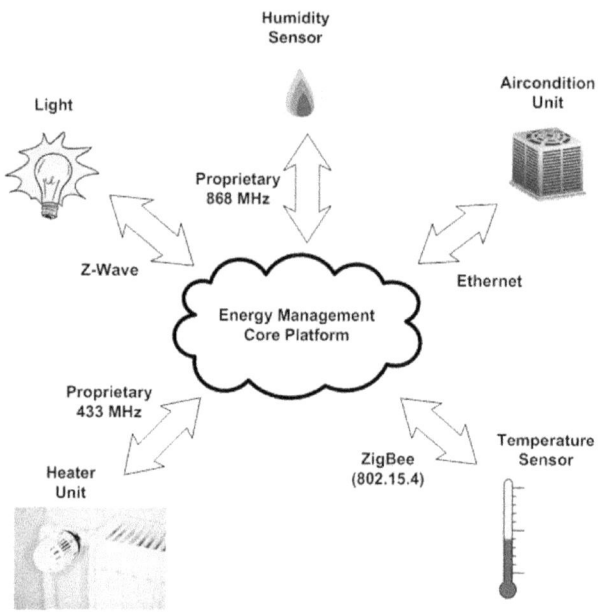

Figure 1 Examples of some of the many different physical link technologies of a classic home automation network to which a central unit must have interface.

- *Interoperability* is a key issue to success. Many different communication technologies are used between HA equipment, see Section 3 below. It is unfortunately unlikely that one network technology will out compete all others in the foreseeable future, and one integrated HA system is easier for the residents to manage and optimize for maximum energy efficiency.
- *Adding a new device* must be easy. HA systems will evolve over time and the residents will want to add new devices to their HA system or replace old malfunctioning devices. On a wired network this is often a simple task, but on wireless networks this is often a challenge. The problem on wireless networks is that there must be a certain procedure to ensure that the device joins the residents HA network and not a neighbours HA network. This procedure is called "pairing of devices", and this task must be easy for the user to accomplish.
- *Battery power* is an issue that influences the choice of wireless protocol used in a HA network. HA systems will often include a number of battery powered devices, e.g. temperature and PIR sensors. To achieve

a decent interval between change of batteries these devices cannot use the traditional Wireless LAN protocol 802.11, but use either a protocol based on 802.15.4 or one of the many proprietary low power network technologies (see Section 3 for more on this issue).
- *The User Interface* is another key issue. Many terms used in HA systems is unknown to most residents, so it is a big challenge to design a user-friendly interface that minimize the efforts necessary to configure a HA system. Some systems go for fully automated HA with no configuration required by the residents, others allow or require everything to be configured, thus demanding a lot of effort from the residents.
- *Cost* is a limiting factor for a widespread use of HA. The associated cost of installation and configuration can be prohibitive. The high cost of the majority of home automation devices is a limiting factor at the moment. Of course general experience from the consumer electronics industry shows that once the quantity of products go up, the prices will decline substantially.
- *Number of nodes* is an issue in larger buildings. Some protocols have a relatively small address space, and can only contain, e.g., 256 devices in a network. And even if the protocol standard defines a huge address space, then the manufacturer sometimes limits the usable address space in their implementation due to memory constrains in the devices.

3 Protocols Used in Home Automation

A typical HA system consists of a central controller that communicates with many distributed sensors and actuators. Most systems use either power line communication or wireless communication, but some systems can use both and some systems use other physical media, e.g. twisted pair cables. On either physical media many different communication protocols are used. In a short investigation we found more than 70 different standard or proprietary protocols were used by different HA systems. This is beyond the scope of this paper, so only a small collection of the protocols used are presented here.

3.1 Power Line Communication Protocols

Power line communication technologies use the household electrical power wiring as a transmission medium, and this enable HA without installation of additional control wiring. Power line communications systems operate by impressing a modulated carrier signal on the wiring system. Since the power wiring system was originally intended for transmission of AC power, the power wire circuits have only a limited ability to carry higher frequencies. The propagation problem and electrical noise are some of the limiting factors for power line communication.

Typically HA power line communication devices operate by modulating in a carrier wave of between 20 and 200 kHz. Each receiver in the system has an address and can be individually commanded by the signals transmitted over the main supply. These devices may be either plugged into regular power outlets, or permanently wired in place.

X10 is an international and open industry standard for communication among electronic devices used for home automation [36]. It primarily uses power line wiring for signalling, where the signals involve brief radio frequency bursts representing digital information, but a wireless radio based protocol is also defined. The protocol was developed in 1975 by Pico Electronics in Scotland, and was the first general purpose home automation network technology and is still used today. Data rates are around 20 bit/s, thus only simple commands like turning devices on and off can be transmitted and it takes approximately 0.75 seconds to transmit a command. Only 256 devices can be addressed in a network.

CEBus (EIA-600) were published in 1992 by Electronic Industries Alliance (EIA) as an enhancement to X10 [7]. CEBus is a set of specification documents which define protocols for products to communicate over power line wire, low voltage twisted pair wire, coax, infrared, wireless, and fiber optics. It uses spread spectrum modulation on the power line. The transmission rate is variable, but the average rate is about 7,500 bits per second. CEBus transmission packets vary in length, depending upon how much data is included. The minimum packet size is 64 bits, which at an average rate will take about 0.009 second to transmit. The standard defines 4 billion device addresses that are set in hardware at the factory.

P1901 is an IEEE standard for broadband communications over power line networks defining medium access control and physical layer specifications and was published December 2010 [16] . The P1901 standard includes two different physical layers (PHY): The OFDM PHY is derived from the HomePlug AV technology and is deployed worldwide in HomePlug-based products. The Wavelet PHY is more narrowly deployed, primarily in Japan.

G.hn is the common name for a home network standard (G.9960/9961) being developed under the International Telecommunication Union (ITU-T) and promoted by the HomeGrid Forum [19]. It supports networking over power lines, phone lines and coaxial cables with data rates up to 1 Gbit/s, but can only address up to 250 nodes in a network [24]. Recommendation G.9960 specifies the Physical Layer and the architecture of G.hn. Recommendation G.9961 specifies the Data Link Layer. G.hn's main focus is broadband communication not home automation but devices in a home automation system may use this protocol.

3.2 Wireless Communication Protocols

The requirements from a HA system to a wireless communication technology is low-cost, low-power, range from 15 to 100 meters and only low transmission rates are needed. These requirements are close to those associated with Wireless Personal Area Networks (WPANs), the main difference is that for a HA network a wider range is preferable. The IEEE has developed a standard to meet these requirements. This standard is IEEE 802.15.4 which was first published in 2003, but is now superseded by IEEE 802.15.4-2006. The later version resolves ambiguities, reduce unnecessary complexity and increasing flexibility in security key usage. The 802.15.4 standard only specifies the physical layer and media access control layer. For the upper layers there are several different opportunities like ZigBee, 6loWPAN or proprietary protocols.

Supported network topologies is a an important issue for a WPAN, as routing can extend the range of the network and minimize the number and extend of radio dead spots that might occur in a home. The three most relevant topologies are star, mesh and tree as shown in Figure 2 and described below:

- *Star*: each device is connected to a central controller with a point-to-point connection. All traffic that traverses the network passes through the central controller. The main advantage of the topology is that it is easy to implement and the primary disadvantage is the lack of routing, which limits the network range to twice the point to point range of the radio front end used.
- *Tree*: a central "root" controller (e.g. the ZigBee Coordinator) is the top level of the tree, and is connected to one or more other devices that are one level lower in the tree (i.e., the second level). Devices on the second level with routing capabilities may have connections to devices on a lower level (i.e., the third level) – and so forth as shown in Figure 2.

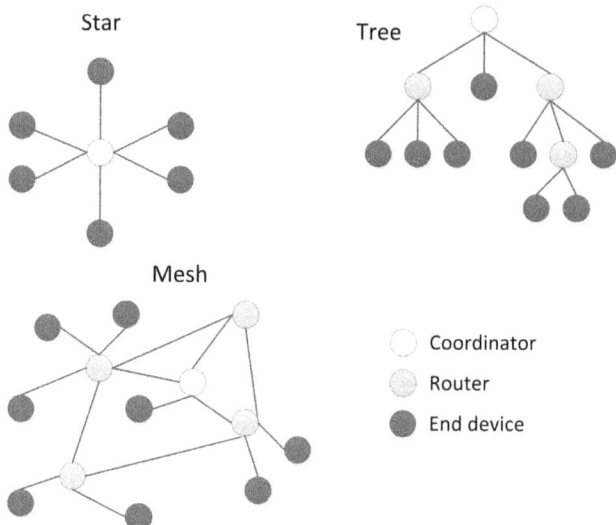

Figure 2 Network topologies.

- *Mesh*: each device must also serve as a router for its neighbour devices. In a network that is based upon a partially connected mesh topology, all of the data that is send between devices in the network takes the shortest path (ideally) between devices. This requires that the devices must perform some type of logical routing algorithm to determine the correct path to use. Well-designed mesh routing algorithms have self-healing capabilities, which make mesh networks very robust.

ZigBee

ZigBee is a low-cost, low-power, wireless networking proprietary standard. The ZigBee Alliance is a group of companies that maintain and publish the ZigBee standard. The term ZigBee is a registered trademark of this group, not a single technical standard. The low cost allows the technology to be widely deployed in wireless control and monitoring applications, the low power-usage allows longer life with smaller batteries, and the protocol supports mesh networking which provides high reliability and larger range. The Zig-Bee specification is based on the IEEE 802.15.4-2003 standard for wireless personal area networks [4]. ZigBee protocols are intended for use in embedded applications requiring low data rates and low power consumption –

individual devices must have a battery life of at least two years to pass ZigBee certification.

The ZigBee stack includes a Stack Profile which can be used to specify the market the protocol will be used in to make better interoperability [31]: e.g. the Home Automation profile is a standard for products enabling smart homes that can control lighting, environment, energy management and security. The Smart Energy profile is a standard for interoperable products that monitor, control, inform and automate the delivery and use of energy and water.

The important device types which makes up the ZigBee network is listed below and in Figure 2 it can be seen how they can be placed in a network of different topologies:

ZigBee coordinator (ZC): The most capable device, the coordinator forms the root of the network tree and might bridge to other networks. There is exactly one ZigBee coordinator in each network since it is the device that started the network originally. It is able to store information about the network, including acting as the Trust Centre and repository for security keys.

ZigBee Router (ZR): As well as running an application function, a router can act as an intermediate router, passing on data from other devices.

ZigBee End Device (ZED): Contains just enough functionality to talk to the parent node (either the coordinator or a router); it cannot relay data from other devices. This relationship allows the node to be asleep a significant amount of the time thereby giving long battery life. A ZED requires the least amount of memory, and therefore can be less expensive to manufacture than a ZR or ZC.

6loWPAN

6loWPAN is an acronym of IPv6 over Low power Wireless Personal Area Networks and it is the name of a working group in IETF [22]. The base specification developed by the work group is RFC 4944 that specifies how IPv6 packets can be sent over IEEE 802.15.4 based networks. The idea of 6LoWPAN is to offer an IPv6-based solution for critical embedded wireless requirement of high reliability and adaptability, long lifetime on limited energy and within highly constrained processing resources to minimize cost. 6loWPAN uses the same RF-chips and low-level protocols as ZigBee, so it have similar specification except the overhead of using IP which reduces throughput. 6loWPAN is an open standard.

Z-Wave

Z-Wave is a proprietary technology developed by the private company Zensys which is the only supplier of the chips implementing the physical layer. The standard is supported by the Z-Wave Alliance which is a consortium of independent manufacturers who have agreed to build wireless home control products based on the Z-Wave standard [3,17,37]. The radio mainly operates in the 900 MHz ISM bands (868 MHz in Europe, 908 MHz in the United States) which often is an advantage, because the 2.4 GHz RF band is typically subject to significant interference due to 802.11 and 802.15.1 devices. The first generations of Z-Wave chips allow transmission at 9.6 and 40 kb/s data, but the recent Z-Wave 400 series supports the 2.4 GHz band and offers bit rates up to 200 kb/s. The transceivers from Zensys allow up to 30 meters indoor range (100 meters outdoors), and the protocol supports the Mesh network topology which enables a wider range, however the address space allows only a maximum of 232 devices in a network. In practice only main powered devices are cable of routing, so a network of only battery powered devices will be unable to route packets.

ONE-NET

ONE-NET is an open-source standard for wireless networking [34]. It is designed for low-cost, low-power control networks for applications such as home automation and sensor networks. It is not tied to any proprietary hardware or software, and can be implemented with a variety of low-cost radio transceivers from a number of different manufacturers. ONE-NET uses UHF ISM radio transceivers and currently operates in the 868 and 915 MHz frequencies but the standard allows for implementation on other frequencies. ONE-NET features a dynamic data rate protocol with a base data rate of 38.4 kbit/s. Indoor range is up to 100 meters (over 500 meters outdoors). Use of mesh mode network topology can extend operational range.

A more detailed description of some of the wireless protocols used in HA networks can be found in [17].

4 The Minimum Configuration – Home Automation Project

In 2008 the Danish Enterprise and Construction Authority decided to fund the project "Minimum Configuration – Home Automation" (MC-HA) with a budget slightly below 1 million Euros. The objective of this project is to develop, through user-driven innovation, a unifying concept of how different electronic solutions can be configured in the home, in such a way that they will become applicable and relevant for users [5]. The partners in the MC-HA project are:

Aarhus School of Engineering	http://ase.iha.dk
The Alexandra Institute	http://www.alexandra.dk
Seluxit A/S	http://www.seluxit.com
Develco Products A/S	http://www.develco.dk

The project is based on participatory design of a multidisciplinary cooperation incorporating user involvement and innovation [30]. Participatory design is a new approach for a number of the involved industrial partners and during the process they learn to use the new methods, while describing the applicability of them for industry value. The project have worked with the following two groups of users:

1. A reference user group consisting of families with or without children
2. Two families who will live in conventional houses in Denmark and Portugal (as living labs).

Based on the user driven innovation, a home automation framework has been created, called the Extendible Protocol Independent unit Controller (EPIC) [35]. Once matured, this framework will be open source for anyone to use. During the development of the initial version of the EPIC framework the following focus points have been given the highest priority:

- *Protocol Interoperability*: The framework must be able to handle several communication protocols in a way that is transparent to the user. This should cover both wireless and power line communication approaches.
- *Solid Backbone*: Since it is the idea that several third party developers will make additions to the framework in the future, it is essential that the framework core is stable. This will ensure that the entire system will not crash due to a faulty addition of new components.
- *Testing Platform*: The user driven innovation created a lot of feedback on how the home automation system shall work. It is important that the framework enables rapid prototype testing of several user interfaces, to support the frequent user feedback. The challenge for the updates of the Graphical User Interface (GUI) comes when the concepts underlying the core components change during the development.

The EPIC platform has been developed and a number of electronic devices (sensors and actuators) have been purchased as commercial products to be integrated in the minimum configuration – home automation setup. Unfortunately (as could have been anticipated) progress on the user interface has not come as far as we had expected because of a large number of interoperability challenges that have turned up with the communication with the different devices. The challenges with different versions of different wireless protocols are described further in Section 5 below.

5 Results Derived from the MC-HA Project

When the MC-HA project started there were no commercial ZigBee devices for home automation on the marked in Europe, but we got hold of a prototype. So the goal for our system was to achieve interoperability between this device and some of the commercial Z-Wave devices for home automation.

We expected the main problems to be concerned with interoperability between the different protocols, but we discovered soon that it was difficult just to setup a functioning net based on only one protocol. In our living labs we made use of both ZigBee and Z-wave devices and we experienced different kinds of challenges with these. The main issues are presented in the subsections below.

5.1 Z-Wave Issues

Regarding the Z-Wave based devices we have encountered the following issues:

- *Network setup*: One of the largest problems for an end user is to setup the network. To do so the user is expected to press a button on the network controller's hardware or software user interface and then press a button on the device that should join the network. The problem is that if the device is already on a network, then it will not join a new network, and simply ignores the press on the button. As a consequence an end user will get no response of what is happening and is likely to assume the device is defect. A new Z-Wave enabled device is not expected to belong to a network when the end user purchases it in a store, but this has been the case with most of the commercial devices we have purchased for the project. From a technical perspective the problem is easy to solve. You remove the device from (any) network. To do so, you just have to select remove device (or something similar) in the network controller's software, and then press a button on the device. The problem is that you need a network sniffer or knowledge about the protocol to figure out what the problem is, and how to solve it.

- *Interoperability*: Interoperability between devices, all using Z-wave, from different manufactures is not as we expected. Devices using the Z-Wave protocol are expected to communicate with each other regardless of who originally manufacturers the devices. However this has unfortunately not been the case in our project. The devices will typical communicate to some extend, but they are rarely fully compatible, and some devices behave strangely when added to a network controller from a different company. One example is a device that would only route packets in one direction, and another device that destroyed the network when added to a network controlled by a controller from a different company.
- *Reachability*: We experienced practical problems with the reachability of the wireless devices in the test houses. Z-Wave devices are supposed to have a range of approximately 100 feet (or 30 meters) in "open air" conditions, but with reduced range indoors depending on building materials. Our experience from the two test houses are that the indoors range is much less. An example is a battery powered sensor placed in the kitchen (device A in Figure 3) that was unable to reach the controller approximately 8 meters away with no concrete walls in between. When we added a switch plug-in (device B in Figure 3) to the network, this device was able to route between device A and the coordinator, so we were able to setup a network that covered every room in the house from one coordinator. But we were surprised that routing was needed for such a short distance.

5.2 ZigBee Issues

Regarding the ZigBee based devices we have encountered the following issue: The first ZigBee specification was ratified in 2004 and made public in 2005, so we expected the ZigBee protocol stack to be mature and stable. However, in our project we found that the protocol stack implementation was not stable yet. This may not be the case for all vendors, but the ZigBee devices used in the project were prototypes from a project partner and this gave us significant additional work because they had to update their firmware frequently. These updates would almost always require changes in our code as

316   P.E. Rovsing et al.

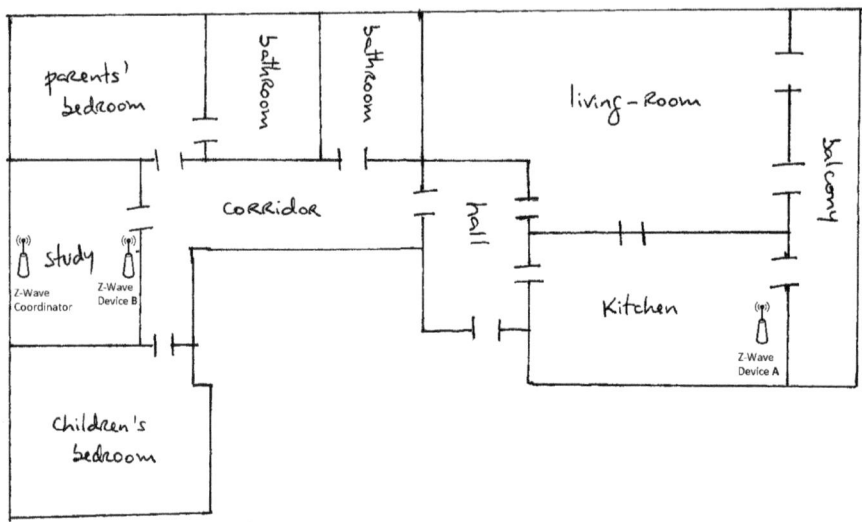

Figure 3 Layout of the Portuguese house (a living lab).

well. One of the major changes was the shift from the ZigBee 2006 protocol stack to the ZigBee 2007 stack release profile 2 (called ZigBee Pro).

5.3 Power Line Communication Issues

Power line communication (a proprietary protocol) was added to our project after we finished the development of our generic communication framework. This was a good test for difficulties in adding support for a new protocol to the framework. It turned out to be rather easy. It only took two weeks to integrate power line communication in our framework, and the power line devices worked very reliably in the test house.

6 Related Work

Currently a number of large European research projects target the use of embedded systems for energy efficient buildings. These range from a project with a cost of more than 17 million Euros such as eDIANA (Embedded Systems for Energy Efficient Buildings) [6], to IntUBE (Intelligent Use of Buildings' Energy Information) [26] and AIM (A novel architecture for modelling, virtualizing and managing the energy consumption of household appliances) [1, 32]. Apart from these projects with their primary focus on embedded systems for energy efficient buildings, there are numerous other projects within the area of development of energy efficient buildings where use of embedded system are only one part of the solution. An example of such a project is "Home for Life" [18]. In addition a number of international research projects across Europe have attempted to solve the problem of interoperability between different link layer technologies by designing a middleware software component making the specific link layer technology transparent for the user. The two largest of these are the Amigo (Ambient Intelligence for the Networked Home Environment) project [14] and the Hydra (Middleware for Networked Devices) project [13].

6.1 eDIANA

The project's main goal is to improve energy efficiency in residential and non-residential buildings through the use of embedded devices. The project is focused on the conceptualization, design, development, demonstration and validation of new devices operating in a uniform platform called eDIANA. The eDIANA Platform is a reference model-based architecture, implemented through an open middleware including specifications, design methods, tools, standards, and procedures for platform validation and verification. The eDIANA Platform will enable the interoperability of heterogeneous devices at the Cell and MacroCell levels, and it will provide the hook to connect the building as a node in the producer/consumer electrical grid. The partners publish many reports at the eDIANA website [11] and links to scientific publications from the project can also be found there. The project has chosen HomePlug 1.0 Power Line Communication technology to be used for providing connectivity between the Cell Device Concentrators (living/working units) and the MacroCell Device Concentrator (buildings). For communication within the Cell the IEEE 802.15.4/ZigBee protocol stack with Home Automation or Smart Energy profiles will be used.

6.2 IntUBE

The project's main goal is to develop tools for measuring and analysing building energy profiles based on user comfort needs. The intend is then to integrate these into Intelligent Building Management Systems to enable real-time monitoring and optimization of energy use. Neighbourhood Management Systems will also be developed to support efficient energy distribution across groups of buildings. These will support timely and optimal energy transfers from building to building based on user needs and requirements. Regarding communication standards IntUBE recommends the use of OPC UA (LonWorks, Konnex), BACnet and oBIX.

6.3 AIM

AIM's main objective is to foster a harmonized technology for profiling and managing the energy consumption of appliances at home. The project will develop a generalized method for managing the power consumption of devices that are either powered on or in stand-by state, and will introduce energy monitoring and management mechanisms in the home network. AIM will also provide services for power distribution network operators (e.g. metering service for energy planning) and for network operators (e.g. remote monitoring and management).

6.4 Home for Life

The goal is to build a sustainable, affordable house that uses readily available technology to negate its imprint on the environment and to promote the health and comfort of its residents [18]. VKR Holding, a private company based in Denmark, is financing a project that will build eight experimental houses in five European countries.

6.5 Amigo

The Amigo (Ambient Intelligence for the Networked Home Environment) project [14] focuses on interoperable middleware aiming at enabling ambient intelligence within the networked home environment by addressing the seamless integration of networked devices and related application services within the home system. The scope is not limited to classic home automation devices, but also aims towards consumer electronics as well as mobile and PC platforms. The Amigo architecture is specifically designed to realize an open networked home system that dynamically integrates heterogeneous devices as they join the network. The dream is to have a home filled with electronics devices that can all "talk to each other" in the future.

Some of the other issues and challenges to the project has attempted to solve is that it is complicated to configure the many different networks within the home automation network. The complexity is high and it is not possible to put too much complexity in the edge-devices due to resource constraints (a strong focus on cost minimization). The architecture has therefore from the beginning been designed and implemented with a service oriented focus. It means that one should no longer think about the washing machine as a device by itself, rather a service provider of a certain functionality (in this case the ability to wash clothes). In this way the network consists of a number of functionalities that work together. By doing this, Amigo facilitates it that the functionalities can interoperate directly. Additionally the Amigo middleware ensures that existing communication protocols can still be used along with new types of communication links. One example is the classic X-10 that Amigo can convert and make a part of the next network.

6.6 Hydra

The Hydra project [13] has from the start aimed at designing, implementing and validating middleware for networked embedded systems that allows developers to develop cost-effective, high-performance ambient intelligence applications for heterogeneous physical devices. The underlying communication layer is considered transparent. It is a tool for further development, targeted towards developers and product manufacturers. The architecture is service oriented like was the case with Amigo. Again Hydra is not targeted specifically towards home automation. Its general structure is equally relevant for other application domains like healthcare, agriculture, etc. Hydra additionally deals with the challenge of developing a framework for secure and trustworthy communication, while at the same time supporting self-adaptive interplay of different components, not only sensors but also controlling components and actuators.

6.7 EnTiMid

The EnTiMid project [23] takes the middleware approach even further. However here the initial setup is based on the idea that one unique and universal middleware is a dream. Therefore to solve this issue, a new generation of schizophrenic middleware in which service access can be generated from an abstract service description has been build. The implementation is called EnTiMid. It is a schizophrenic middleware which supports various services access models (several personalities): SOAP (Simple Object Access Protocol), UPnP and DPWS (Device Profile for Web Services). The model personalities are generated using a model driven engineering approach.

7 Future Tendencies

The consumer is interested in interoperability, ease of use, configuration and installation and the overall cost of a HA system. The current solutions that combine multiple areas of HA such as an alarm system and a heating control system, may be easy enough to use, but these systems still require a relatively highly skilled person for the installation and the configuration. Furthermore in many systems the home owner will not be able to adjust the system to her or his requirements without consulting an expensive expert. A good solution to the usability problem would thus also solve the cost problem up to a point. In the past companies such as Apple, Nokia and Microsoft have made complex technology more accessible to the wider public. The same leap in usability is needed for the widespread acceptance of home automation solutions. As there is currently increasing activity in the development of such solutions it is only a matter of time until the right solution emerges.

In other domains there have been a strong movement towards global standards. It would be very difficult to get a decent market share for a wireless interface card for labtop computers which was incompatible with the WiFi specification, or a hand free head set for a cell phone incompatible with Bluetooth. But even standards within the field of home automation exist, there is no clear movement towards them. A study by IMS Research, "The World Market for Low-power Wireless 2011 Edition" [27] finds that, in 2009, of the 20 million IEEE 802.15.4 ICs shipped, fewer than half were ZigBee certified. And within home automation 802.15.4 is only one of the many different protocols used. One of the reasons why this is the case is due to the number of standards. There are too many competing standards and no obvious winner among them. Another reason may be the structure of the market. In the home automation marketplace there are many small national companies compared to the cell phone marketplace which is dominated by large multinational companies.

Currently there is no clear winning standard among the proprietary and open standards available. In the past the winning standards in other areas had a good open specification. These standards include TCP/IP, USB, HTML and XML. An open standard without the requirement of certification and the payment of royalties to some part of the industry can form the basis for a competitive an successful market introduction of home automation.

A winning home automation solution will probably comprise multiple standards both wired and wireless. Furthermore such a solution will probably include use of proven technology such as 802.11 or bluetooth as this will

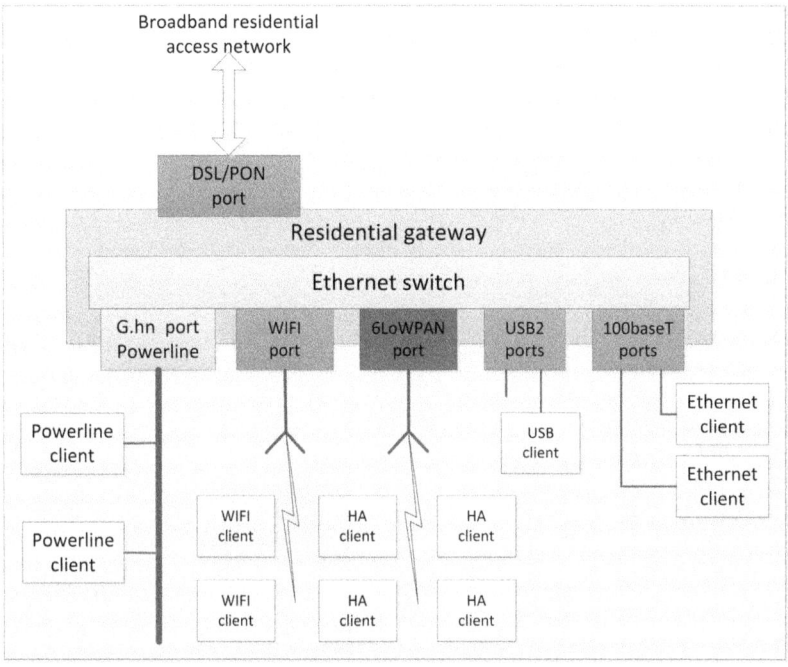

Figure 4 Vision for a HN topology that suits Home Automation.

make the integration into the existing infrastructure easier. The interoperability can be achieved by using proven technologies such as IPv4 or IPv6 based protocols. Oksman and Galli [24] has a vision of one integrated infrastructure for the home. They promote G.hn and figure 2 in their paper [24] shows an example of Home Network (HN) topology associated with residential access. But they miss one important issue to fully connect everything. Many home automation systems include several battery powered devices such as temperature sensors. They need a wireless low-power protocol. To integrate smoothly into the other protocols used this could be 6LowPAN over 802.15.4 as shown in Figure 4. Then IP could be the protocol to integrate the different network technologies.

Depending on the specific area where the solution is used, be it automatic meter reading or an alarm system, different solutions will probably become the de facto standards. Therefore middleware solutions that can seamlessly bridge between multiple standards are the only option for the foreseeable future as long as there is no clear winning standard in the home automation market. On power line there are two competing standards: G.hn and IEEE's

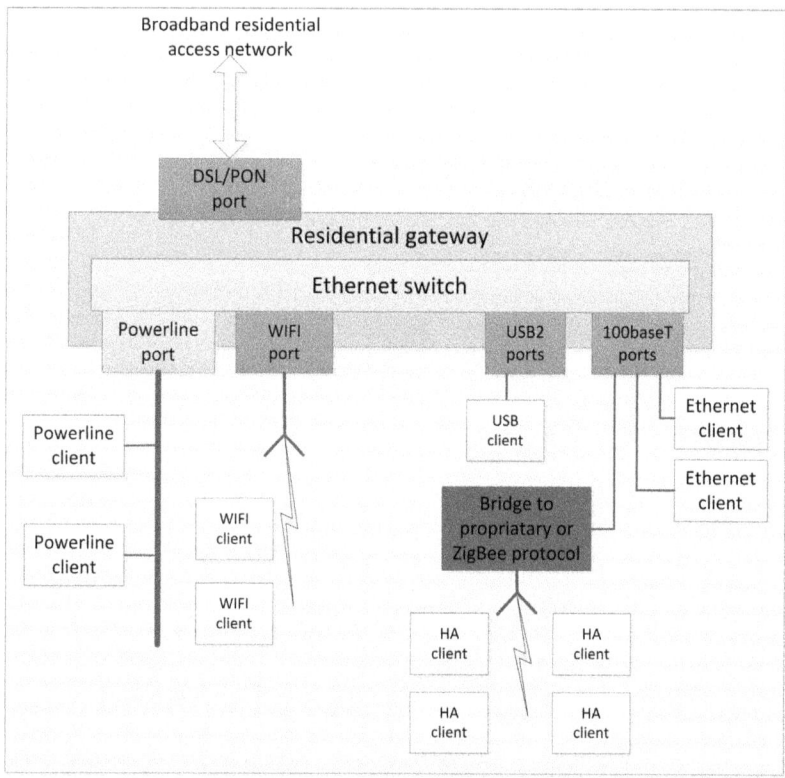

Figure 5 A more realistic vision for a HN topology that include Home Automation.

P1901 and only time can tell if one of them will win the market and leave the other and all the proprietary protocols behind. On the wireless media there are many alternatives to the ZigBee standard. Even if ZigBee has better performance than most of the competing protocols (e.g. bandwidth and range), this does not necessarily make it a winner. In the beginning of 2011 there are still relatively few (25) ZigBee certified products for home automation [2]. Therefore a more realistic HN topology is the one shown in Figure 5.

Furthermore a good open application layer protocol for home automation is needed. Such a protocol might be based on XML or the more suitable EXI standard defined by the W3C [33].

8 Concluding Remarks

Despite the promising potential for using intelligent home automation and improving the energy consumption in private homes [29], it seems that the maturity of the electronic devices to be used in such networks are not yet where they need to be for this to really have any measurable effect. To achieve a measurable carbon emission reduction from home automation technologies the technology has to be extremely widespread. It must be employed in a significant part of the private households, i.e. it must be extremely easily employable. It is amazing how difficult a simple task such as adding a new device to a wireless network can be for an end user to accomplish. In our view it is disappointing that it is so difficult to realize a heterogeneous network of home automation devices that can interoperate appropriately. We believe that in order to remedy this situation both the industry for wireless home automation electronics as well as the standards institutions for wireless technologies quickly needs to create sufficient standards, preferably with backward compatible interoperating protocols. For the power line some of this is being done by ITU through the G.hn standard, promoted through the HomeGrid Forum [15].

One way of overcoming some of the challenges we have found in the MC-HA project is to move wireless home automation (actually both wireless and fixed network parts) into a pure embedded Internet setup [21]. This was initially proposed already in 2001 by Finch [12], but it was rejected since it was considered a far to heavy-weight type of solution. However, major advances in sensor networks as well as small scale sensor electronics have changed this situation significantly. With technologies such as the IETF 6LoWPAN [22], it is now feasible to run an IPv6 solution even on an 8-bit micro controller [10,20]. On top of that, the "upper" layers like the web server are being adapted to very resource constrained platforms [9]. With such an approach IPv6 will offer one single network layer that can cover all aspects. Additionally, different link layer technologies can be used in the same home network at the same time, wireless as well as fixed.

Acknowledgements

We would like to thank the Danish Enterprise and Construction Authority for financial support for this project and all our partners in the project for their active participation.

References

[1] AIM – A Novel Architecture for Modelling, Virtualising and Managing the Energy Consumption of Household Appliances. http://www.ict-aim.eu/.
[2] The ZigBee Alliance. ZigBee home automation certified products. http://www.zigbee.org/Products/CertifiedProducts/ZigBeeHomeAutomation.aspx, 2011.
[3] Z-Wave Alliance. Z-wave: Products that speak z-wave work together better. http://www.z-wave.com, December 2009.
[4] ZigBee Alliance. *ZigBee Specification*. January 2008.
[5] Minimum Configuration Home Automation. http://iha.dk/minimumconfiguration.
[6] Chiara Buratti, Andrea Conti, Davide Dardari, and Roberto Verdone. An overview on wireless sensor networks technology and evolution. *Sensors*, 9:6869–6896, 2009.
[7] Cebus Protocol. http://en.wikipedia.org/wiki/CEBus.
[8] S. Darby. Making it obvious: Designing feedback into energy consumption. In Paolo Bertoldi, Andrea Ricci, and Anibal De Almeida (Eds.), *Energy Efficiency in Household Appliances and Lighting*, pp. 685–696. Springer Verlag, 2001.
[9] S. Duquennoy, G. Grimaud, and J.-J. Vandewalle. SMEWS: Smart and mobile embedded web server. In *Proceedings of IEEE International Conference on Complex, Intelligent and Software Intensive Systems*, March 2009.
[10] M. Durvy, P. Wetterwald, B. Leverett, e. Gnoske, M. Vidales, G. Mulligan, N. Tsiftes, N. Finne, and A. Dunkels. Poster abstract: Making sensor networks IPv6 ready. In *Proceedings of the 6th ACM Conference on Embedded Networked Sensor Systems*, November 2008.
[11] eDIANA Embedded Systems for Energy Efficient Buildings. http://www.artemis-ediana.eu/index.php.
[12] E. Finch. Is IP everywhere the way ahead for building automation? *Facilities*, 19(11):396–403, 2001.
[13] Middleware for Networked Devices HYDRA Project. http://www.hydramiddleware.eu.
[14] Ambient Intelligence for the Networked Home Environment, Amigo Project. http://www.hitech-projects.com/euprojects/amigo/.
[15] The HomeGrid Forum. http://homegridforum.org/.
[16] Shmuel Goldfisher and Shinji Tanabe. IEEE 1901 access system: An overview of its uniqueness and motivation. *IEEE Communications Magazine*, 48:150–157, October 2010.
[17] Carles Gomez and Josep Paradells. Wireless Home Automation Networks: A Survey of Architectures and Technologies. *IEEE Communications Magazine*, 48:92–101, June 2010.
[18] Ellen Kathrine Hansen. Home, Smart Home. *IEEE Spectrum*, 47(8):34–38, August 2010.
[19] Cebus protocol. http://www.homegridforum.org.
[20] J. Hui and D. Culler. IP is dead, long live IP for wireless sensor networks. In *Proceedings of the 6th ACM Conference on Embedded Networked Sensor Systems*, November 2008.
[21] Matthias Kovatsch, Markus Weiss, and Dominique Guinard. Embedding internet technology for home automation. In *Proceedings of the 15th IEEE International Conference on Emerging Technologies and Factory Automation*, September 2010.

[22] N. Kushalnagar, G. Montenegro, and C. Schumacher. RFC-4919, IPv6 over low-power wireless personal area networks (6LoWPANs): Overview, assumptions, problem statement and goals. IETF Request for Comment, 2007.
[23] Grégory Nain, Erwan Daubert, Olivier Barais, and Jean-Marc Jézéquel. Using MDE to build a schizophrenic middleware for home/building automation. In *ServiceWave'08: Proceedings of the 1st European Conference on Towards a Service-Based Internet*, pp. 49–61. Springer-Verlag, Berlin/Heidelberg, 2008.
[24] Vladimir Oksman and Stefano Galli. G.hn: The new ITU-T home networking standard. *IEEE Communications Magazine*, 138–145, October 2009.
[25] Wouter Poortinga, Linda Steg, Charles Vlek, and Gerwin Wiersma. Household preferences for energy-saving measures: A conjoint analysis. *Journal of Economic Psychology*, 24:49–64, 2003.
[26] The IntUBE Project. http://www.intube.eu.
[27] IMS Research. The World Market for Low-power Wireless, 2011 Edition. http://imsresearch.com/news-events/press-template.php?pr_id=1807, December 2010.
[28] John Rohde, Sune Wolff, Thomas Skjødeberg Toftegaard, Peter Gorm Larsen, Kenneth Lausdahl, Augusto Ribeiro, and Poul Ejnar Rovsing. Optimizing energy usage in private households. In R. Prasad, S. Ohmori, and D. Šimunić (Eds.), *Towards Green ICT*, pp. 185–209. River Publishers, 2010.
[29] John Rohde, Sune Wolff, and Thomas Skjødeberg Toftegaard. Strategies for releasing the green potential in home automation. In *Proceedings of WPMC'09: The 12th International Symposium on Wireless Personal Multimedia Communications*, September 2009.
[30] Douglas Schuler and Aki Namioka (Eds.). *Participatory Design: Principles and Practices*. Lawrence Erlbaum Associates, 1993.
[31] MeshNetics Team. Meshnetics: Easy wireless for things. http://www.meshnetics.com, December 2009.
[32] S. Tompros, N. Mouratidis, M. Draaijer, A. Foglar, and H. Hrasnica. Enabling applicability of energy saving applications on the appliances of the home environment. *IEEE Network*, 23(6):8–16, November/December 2009.
[33] World Wide Web Consortium (W3C). Efficient XML Interchange (EXI) Working Group. http://www.w3.org/XML/EXI/, January 2011.
[34] ONE-NET web site. FAQ, http://www.one-net.info, 2011.
[35] Sune Wolff, Peter Gorm Larsen, Kenneth Lausdahl, Augusto Ribeiro, and Thomas Skjødeberg Toftegaard. Facilitating home automation through wireless protocol interoperability. In *Proceedings of WPMC'09: The 12th International Symposium on Wireless Personal Multimedia Communications*, September 2009.
[36] X10 protocol. http://en.wikipedia.org/wiki/X10_(industry_standard).
[37] Zensys. *Z-Wave Protocol Overview*. April 2009.

Biographies

Poul Ejnar Rovsing works as an associate professor at the Engineering College of Aarhus, Denmark. After receiving his B.Sc. degree at the

Engineering College of Aarhus in Electronic and Software Engineering in 1987, he worked a few years in the ICT industry. He returned soon to academia to teach different topics within software engineering. During most of the 1990s he combined the teaching with a part time work for the Hemodynamic Research group at Skejby Sygehus, Aarhus University Hospital. During this period he was co-author on a few papers. After a period of only teaching, he has taken up research again, but now within the field of home automation technologies.

Peter Gorm Larsen is currently a professor at Aarhus School of Engineering where he act as the team lead for the software engineering team. After receiving his M.Sc. degree at the Technical University of Denmark in Electronic Engineering and Computer Science in 1988, he went to industry to bridge the gap between academia and industry. He later returned and did an industrial Ph.D. degree which was completed in 1995. He gave industrial courses all over the world, and had an industrial career until he decided to return to academia in 2005. His prime research interest is to improve the development of complex missing critical applications with well-founded technologies. He is the author of more than 70 papers published in journals, books and conference proceedings and a couple of books.

Thomas Skjødeberg Toftegaard (former Thomas Toftegaard Nielsen) holds a M.Sc.E.E. (1995) and a Ph.D. (1999) in wireless communications from Aalborg University, Denmark. On 1 April 2009 he was appointed Professor in Communication Technology at Aarhus School of Engineering, Aarhus University. Additionally he serves as Director of the Electrical Engineering and Information and Communication Technology group at Engineering College of Aarhus. Professor Toftegaard is affiliated to the Computer Science Department at Aarhus University where he is a member of the research committee. His main research interests is on future intelligent wireless connectivity, working with mobile communications, sensor networks, massive dense network architectures, network complexity, wireless IP, software defined radio, cognitive radio and ubiquitous wireless networks. In addition to academic work, his professional carrier includes an over 14 years tenure in the high-tech industry in the Scandinavian region. With responsibilities exclusively in the R&D, project management and product development domains, he has developed expertise in the design and development of mobile communication systems, mobile phones and product components. Professor Toftegaard has published one book on wireless communications, a book chapter on Wireless IP as well as a number of journal and conference papers.

Daniel Lux is the co-owner of Seluxit which was founded in 2006. He studied computer science at the University of Groningen in the Netherlands. At Seluxit he has been involved in the development of a middleware platform for home automation based on an open XML standard and has developed different kinds of home automation products.

Energy Efficiency as Input for Context-Aware Group-Based Communications

Nuno Coutinho[1], Tiago Condeixa[1], Susana Sargento[1] and Augusto Neto[1,2]

[1]*Instituto de Telecomunicações, University of Aveiro, 3810-193 Aveiro, Portugal; e-mail: {nunocoutinho,tscondeixa,susana}@ua.pt*
[2]*Instituto de Informática, Universidade Federal de Goiás, 74001-970 Goiânia, Brazil; e-mail: augusto@inf.ufg.br*

Received 31 January 2011; Accepted: 15 March 2011

Abstract

Heterogeneity will prevail in Future Networks at many technology levels, including devices, access networks, services and applications. Moreover, the exponential proliferation of mobile computing has been followed by concerns and research efforts regarding a sustainable development of wireless systems, specially related with energy efficiency. The envisioned future networking environments will be able to build a very rich source of information that can be wisely used, not only to improve users' service perception, but also to enhance network performance at different perspectives (e.g., quality of service, resource management and energy consumption). We believe that this is the main reason behind the increasingly popular concept of context-awareness. Current context-driven architectures aim to provide a platform to enable personalized services and smart networking, adapting contents and the way they are delivered to the users according to their particular context (device, available energy and preferences) and the network and environment context (location, technologies and energy consumed). However, given the proliferation of group-based communications and consequent traffic demands, we show in this paper how context information, including energy-related context, may influence content delivery to users, employing the information available to create groups of users according to their common capabilities. This way, we aim to achieve a higher level of personalization in multiparty services, improving users' quality of experience. Through a service-aware group-oriented content delivery framework, we employ algorithms for core and access network selection that are flexible enough to be guided by any constraint. The developed selection scheme uses context information from user, network and environment to decide, among the available networks, the most suitable one regarding its communications needs and optimizing network constraints, such as quality of service provided and energy required. The outcomes obtained through the simulation of the proposed framework show that context-based grouping and network selection are able to improve the users' quality of experience.

Keywords: green computing, multicast, quality of service, overlay, context-awareness, heterogeneity.

1 Introduction

The Information and Communication Technologies (ICT) have moved beyond their traditional areas of application to form smart spaces, scenarios harmonizing a number of fixed and mobile networking technologies and providing users with best possible personalized experience. In smart spaces, ICTs help to address various challenges of the modern society, starting from efficient indoor use of energy to intelligent logistics and broad-scope deployment of the green principles and technologies. Despite notorious developments over the last decade, ICT providers have always been struggling to minimize their overall energy consumption, mainly due to the magnitude expected for the next years, massive data rates of thousands networking sources distributed over wide-areas, leading to higher energy consumption.

Initially, the energy concerns were related with terminal autonomy and economic benefits. Nowadays, however, the global urgency to investigate new technologies that can enable a more sustainable society and reduce their environmental impact has raised a new research trend, commonly referred to as "Green Computing". Thus, the pursuit for novel mechanisms that may reduce, not only their ecological footprint but also the operational costs, is crucial for designing Future Networks (FNs).

In FNs, heterogeneity will prevail as the main feature of future networking environments, as depicted in Figure 1. The diversity of facilities

Energy Efficiency as Input for Context-Aware Group-Based Communications

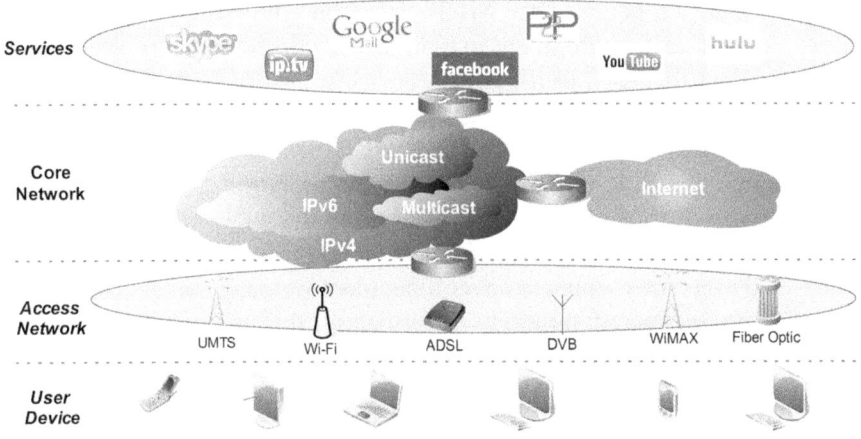

Figure 1 NGN heterogeneous environment.

may be found at different levels, ranging from the terminals to the services: user devices have distinct processing capabilities and mobility facilities; the panoply of access network technologies differ on bandwidth, coverage and energy required; the co-existence of different transport technologies in the core and the variety of application requirements. It is a common understanding that this diversity represents, more than a challenge, an opportunity to differentiate the services provided and, as consequence, user experience. In this sense, context-awareness concept has been prominent in the last years within the research community, aiming to endow future architectures with the ability to sense surrounding conditions so that taking advantage of that information towards innovative network management. Therefore, novel architectures guided by context have been proposed, focusing on adaptable content delivery according to user capabilities and Quality of Service (QoS) requirements.

The recent growth of group-based multimedia sessions, such as IPTV, poses new challenges, mainly due to their bandwidth-intensive and time-sensitive demands. Recently, multicast technology has gained a new stimulus as the most suitable solution to address bandwidth-constrained demands. Nevertheless, operators remain reluctant regarding multicast scalability, despite recent efforts to overcome this issue. Moreover, legacy multicast is not prepared to handle end-to-end multiparty sessions, which encompass one-to-one or one-to-many channels crossing interconnected networks with different underlying transport technologies (i.e., unicast/multicast, IPv4/v6, etc.). To that, we believe that wisely using the knowledge available about users, network and environment may influence content delivery according to any constraint (energy-, user-, session- or network-oriented). Thus, context-awareness is a key feature in FNs architectures and an added value for service providers, which may differentiate and enrich their offer.

In terms of end-to-end heterogeneity, the design of a context-aware architecture needs to be wisely performed, given the wide range of constraints that can be used to guide the selection of best networks to keep required nodes and users interconnected over time; moreover, any context change may trigger network reconfiguration. Nevertheless, this motivates the research community to develop new solutions in order to better manage network operator resources while meeting the QoS requirements of applications and user preferences.

Service personalization may be an added value for operators, although being a challenging goal considering group communication environments. Thus, FNs must be able to wisely contextualize available information and provide the services to users' needs, capabilities and preferences. Nevertheless, context-aware networks face several challenges: from one side, context (e.g. preferences, location, mobility, resources, energy) may be used to guide the selection of best networks for the groups of services (respecting the energy-efficiency requirements); from the other side, any context change may require a complete restructuring of the network and its sessions, which is even more difficult and challenging given the heterogeneous cooperating technologies behind FNs.

The aim of our work is to design an architecture to efficiently support multiparty content delivery. The main idea concerns hiding network heterogeneity, and taking advantage of this heterogeneity in the provision of more personalized multiparty sessions. Thus, in order to capitalize the diversity of FNs, we focus on the deployment of intelligent network selection mechanisms fed by sensed context information. The purpose of these mechanisms is to select the networks that best fit communication requirements, and also adapt the content distribution to current network, user, session and surrounding environment conditions. This process is of an added difficulty in group communications, considering the variety of user context, making difficult the provision of a unique multicast session that may fulfil several user requirements simultaneously.

In this sense, we adopt the grouping strategy, dividing users according to their context, such as terminal capabilities, network conditions, location, etc. The work addressed in this paper focuses on the access and core intelligent selection driven by contextualized capabilities from the network- and user-side, as well as the specification of an efficient mechanism of grouping/sub-grouping of mobile users supporting network/user context. Thus, our goal is to increase the experience perceived by the user, offering context-aware multiparty services, while optimizing network revenues and energy constraints. The outcomes of the simulated approach show that context-awareness improves the overall network performance while providing a personalized multiparty content delivery.

The remainder of this paper is organized as follows. The related work is briefly presented in Section 2. Section 3 describes the architecture, functions and mechanisms. Section 4 details the selection algorithms developed and their evaluation through simulation. Finally, Section 6 concludes the paper and Section 7 provides an insight into possible research directions.

2 Related Work

Nowadays, one may find in the literature a strong research effort addressing the challenges of FNs regarding their environmental impact and energy efficiency. There are many reasons fuelling the investigation towards so called "green computing", e.g., economic, marketing, and environmental. A good and ambitious example of the efforts being made on this topic is the global consortium GreenTouch [9]. Organized by Bell Laboratories, its goal is to create the technologies that are needed to make communication networks 1000 times more energy-efficient than they are today. Focused more on the radio access networks, the European research project EARTH [8] also aims to develop novel concepts and solutions that may reduce energy consumption by 50%.

Despite this current trend in green communications, network energy issues have been tackled since the beginning of mobile communications, focusing on base stations energy consumption and mobile devices autonomy. However, given the exponential growth of mobile terminals and the global concerns about a sustainable society, these energy issues become even more relevant and urgent to solve. Thus, the research community has been pointing out possible improvements at the different layers of the communication protocol (physical, network and applications level), as in [5, 6].

Wired networking has also been target of the *greening* process, as presented in the study in [2]. This survey provides a taxonomy of the related work identifying the main reasons for the current state of energy inefficiency. Moreover, due to the fast growing of multiparty multimedia applications (e.g.

IPTV) and their increasing bandwidth demands, it raises the necessity to increase the efficiency of resource and energy management. As stated in [20], the power consumption for high access rate is higher in the core than in the access networks, which requires novel concepts and energy-oriented routing protocols and network design as proposed in [4].

Another hot topic in the literature nowadays is the concept of context-awareness. It was originally introduced in computing environments by Schilit [22], and years later defined in a broader sense by Dey and Abowd [7] as *"any information that can be used to characterise the situation of an entity"*. The entity may be any participant considered relevant in the interaction between user and service, the user itself, surrounding conditions, place or person. The perception of this diversity of information is the first step towards the development of network architectures sensible to their surrounding conditions. Thereby, we believe that context-aware approaches can be used to improve the efficiency of the communication at different levels, such as resource management or energy waste. The challenge is to choose the correct constraints according to a pre-defined optimization goal, and develop algorithms and mechanisms that can materialize the information sensed towards that objective.

Considering the popularity and potential of this concept, there are many proposals in the related work addressing mechanisms for context gathering, modelling and reasoning [10, 23]. A virtual network operator supporting context-aware services is proposed in [21], using information about location, device and identity management across multiple access networks, abstracting underlying technologies. Within the scope of the Daidalos project, Williams et al. [24] introduce the Context Management Subsystem to collect and store environment knowledge through sensors, which is later used by the personalization module interacting with a context broker to perform the adaptations in the service provided. However, given the vague definition of context and considering that context-aware services are dependent of the information used, it is important to assess the quality of context itself, as suggested in [3].

Nevertheless, these proposals are too focused on using context in a very user-centric way, abstracting from network conditions and what they may offer. A more comprehensive architecture is proposed by Mathieu et al. [16], introducing a new level of network adaptability based on self-management schemes to address the high dynamics of such environments, using overlay networks that try to fulfil service requirements and adapt content delivery to user and network context.

The heterogeneity of FNs and mobile devices allowed one to envision services as the ones referred to in [11,12]. Under this subject, many proposals may be found in the literature aiming to provide an intelligent access network selection [13, 15, 19]. Farooq et al. [1] provide an access selection solution using the Technique for Order Preference by Similarity to Ideal Solution (TOPSIS). It uses a matrix that models all the access networks regarding the attributes assigned to them. Then, for each access network, the distances for the best and worst case are measured using Euclidean distances, establishing a ranking. A more comprehensive selection scheme is proposed by Xing et al. [25], which maps all traffic flows through the available access networks in order to accomplish as much as possible the following requirements: satisfy user preferences, maximize the number of traffic flows admitted and minimize power consumption cost while satisfying QoS needs. It uses a variant of the bin-packing problem, adopting algorithms derived from the First Fit Decreasing algorithm, in order to obtain near-optimal solutions and reduce computational effort.

The related work study how that these approaches have the common purpose of selecting the access network that best meets communication needs taking into account network conditions and user preferences. However, our analysis attest that different parameters are employed in each of these schemes, dealing only with unicast traffic and not considering constraints related to multiparty sessions.

By definition, Quality of Experience (QoE) [14] refers to the overall acceptability of an application or service, as perceived subjectively by the end-user. Thus, it is affected by the quality and efficiency of each actor of a communication: source, content quality, terminal, network and user. Thereby, we support that to achieve higher values of QoE (regardless the metrics), the most suitable strategy is to involve simultaneously network and user information in the decision processes. However, given the diversity of information and constraints involved one may defined the optimization goal: user-, network- or energy-oriented.

3 Architecture

The architecture proposed aims essentially to enable effective multiparty content delivery in heterogeneous networking environments, mainly regarding multiparty mobile multimedia applications. It is our goal not only to cope with FNs heterogeneity, but also to take advantage of it in order to enhance user experience. Energy requirements and constraints are included both as

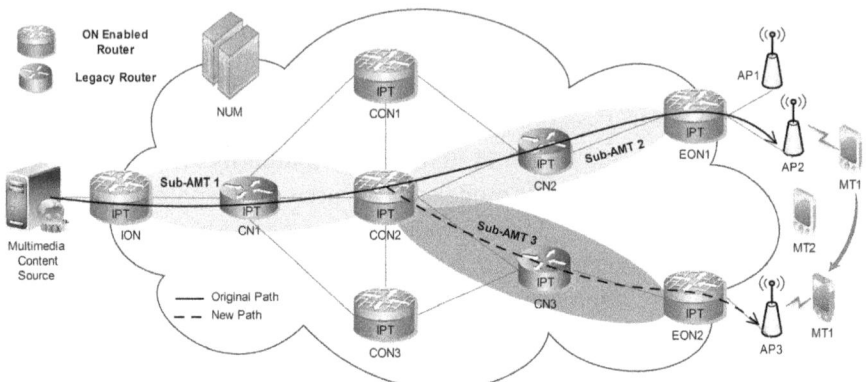

Figure 2 Multiparty content delivery architecture.

part of the network and user context. In the envisioned architecture, any change of context may influence established sessions, requiring their renegotiation and, consequently, reconfiguration of the network. Moreover, mobility will add a new dynamic to FNs, raising the necessity to efficiently deal with constant changing environments.

To accomplish this, we propose a context-driven architecture supported by the following features: (i) context-awareness, (ii) session management, and (iii) multiparty content delivery. The context acquisition framework is built through several sensors distributed throughout the network, which feed a central repository named Context Broker (CB). This feature is crucial to provide the knowledge for the development and support of personalized services, adapting contents and the way they are delivered according to the user particular context. At the session layer, the collected information is used to manage session events and create the session context, consisting of session's group members identification and QoS requirements (bandwidth, delay, loss, etc.).

In this paper, we focus on the Context-Aware Multiparty Transport Framework, represented in Figure 2, which comprehends a set of elements and concepts to control in an hierarchical way the content delivery and network resources, considering scalability concerns that usually come along with context-sensitive architectures.

Network Use Management (NUM)

NUM represents the higher level of control that makes use of the context information available to proper react to context changes or events. It provides smart context-aware network selection (for access and core), combining user, environment, session and network related information. Core network selection is responsible to determine appropriate data paths for the content delivery tree in the core network (described in next section). Moreover, NUM is also responsible for maintaining the mobile terminals connected to the access networks that best meet their communications needs. Through an access selection algorithm, it offers the best combination of users and services throughout a heterogeneous access network system. Consequently, it is expected to achieve a more uniform distribution of the load between the different radio access technologies and core nodes, while satisfying users' requirements and preferences. NUM's decisions are enforced in a distributed manner taking advantage of the second level of control of the architecture, as detailed below.

Abstract Multiparty Trees (AMT)

The concept of abstract transport multiparty trees relies on an overlay network that operates on top of the IP network layer. This concept allows the general transport in multicast trees hiding network dynamics and underlying transport technology heterogeneity (i.e., Unicast/multicast, IPv4/v6, etc.). This feature is particularly important regarding the lack of deployment of multicast and IPv6 in certain network segments. This concept allows the coexistence of these technologies with legacy ones, enabling an incremental deployment of FNs. Thus, AMTs allow end-to-end multicast content transport over heterogeneous network segments, also providing independence between source and listener trees, easing local reconfiguration processes.

These local network segments are called sub-AMTs, see Figure 2, where all nodes (edge and core) composing a sub-AMT implement the same transport technology. NUM coordinates the edges of each sub-AMT, the Overlay Nodes (ONs), which may be any core node embedded with more resources and functionalities. Following this strategy, the processes of resource allocation and network reconfigurations are of decreased complexity, since resources and QoS are assigned per sub-AMT. Moreover, since the AMT is divided in several sub-AMTs, local reconfigurations can be done inside each sub-AMT, without changing ONs. This process does not involve NUM or other sub-AMTs, increasing scalability and easing the reconfiguration process.

Internet Protocol Transport (IPT)

The IPT component is embedded on each network node, being responsible for network resource allocation, setting sub-AMTs between ONs. IPT is based on the existing Multi-Service Resource Allocation (MIRA) [17] proposal to support scalable per-class QoS provisioning associated with IP multicast control. MIRA assures the quality level requested for each flow of a multiparty session, by adjusting the resources of selected Class of Service (CoS). The CoS bandwidth is controlled following an ingress-to-egress approach in the direction of the access-router of the receiver. In an egress-router, the selected CoSs are configured taking into account the Service Level Agreements (SLA) established with the neighbor network. Then, PIM-SSM is triggered. In IP multicast-enabled Sub-AMTs, MIRA correctly supplies the Multicast Routing Information Base (MRIB) during the CoS resource adjustment (ingress-to-egress), aiming at guiding PIM-SSM messages to install the referred multicast state along the QoS path. IPT supports a distributed per-class resource control, thus whereas session establishment can be requested in a per-flow basis, resources are configured per-aggregate. IPT provides seamless resilience operations, since it attempts to reconfigure multicast trees between ONs inside the sub-AMTs, without changing multicast groups address and ONs.

3.1 Context

A context-aware architecture must contain three key procedures (see Figure 3) in order to adapt the service provision according to surrounding conditions: context acquisition, context modelling and context reasoning. In our work we do not address the first two mechanisms, since these are included in the Context Acquisition Framework. However, context gathering and modelling is crucial in order to enable the architecture with means to obtain knowledge about different network entities. Moreover, as this kind of information is not in a suitable format to be reasoned by intelligent schemes, it is critical that all the context information gathered be modelled or formatted.

Figure 4 depicts several sources of context from different communication actors: service, user, device and network. Within each of these entities, there are several characteristics that may be used with distinct purposes:

- *Application*: requirements quantified in QoS metrics (bandwidth, packet loss, delay);
- *User*: the user as a person has its own preferences and profile, including energy constraints, which can be used to improve its QoE. Location, trajectory and speed can also be used by proactive and predictive algorithms;

Energy Efficiency as Input for Context-Aware Group-Based Communications 339

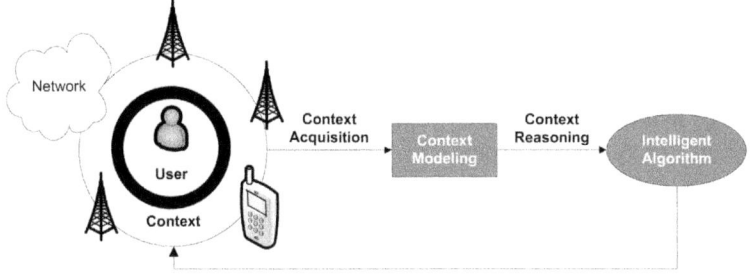

Figure 3 Context-awareness cycle.

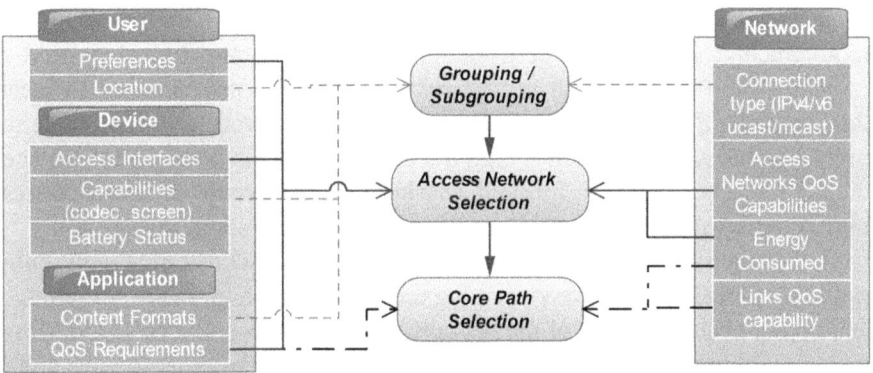

Figure 4 Context-aware mechanisms.

- *Device*: user's terminal related characteristics such as display resolution, available interfaces and battery status;
- *Network*: network resources, technology, energy requirements, performance feedback, reliability, operator policies and cost;
- *Environment*: noise, weather, type of building.

3.1.1 Sub-Grouping

Since the architecture is targeting multiparty communications, the integration of this kind of services and personalization is a very challenging task. In this work, we pursue the balance between these two conflicting approaches, introducing the concept of grouping and sub-grouping. Although the same content is sent to all group members, its delivery needs to be adapted for each user based on his particular context. Considering the variety of context information available, adaptations are performed at Session and Network levels based on the context of the respective layers. At the session level, users can be arranged according to their preferences and/or device capabilities, such as media codecs, resolutions supported, location, battery status or expected energy consumed using a certain access. In case of on-going sessions, the grouping mechanism must also react if any relevant context change occurs (handovers, QoS degradation or device mobility), adapting user's session accordingly.

However, here our focus is on the sub-grouping at the network level, where we consider two different adaptation processes that may set different sub-groups: access technology and network resources. The former is based on aggregating users that share the access that best fit their communication needs. Concerning the network resources, each group will have different QoS requirements. Thus, at the network level, each sub-group has associated a multicast delivery tree with proper network resource provisioning to meet the demands of the codecs used. By adopting this approach, it is possible to reduce global re-arrangements, e.g., if a user changes his access network, and the QoS levels cannot be fulfilled within the same group, different codecs may be adopted and the user is transferred to a more suitable sub-group.

Given the users' heterogeneity, including too many constraints in the sub-grouping process can be greedy, taking the risk of obtaining many sub-groups and do not take advantage of the network resources that could be saved using groups with more users. A fair compromise is required in order to have a fair as possible personalized multiparty communication.

3.2 Selection

Selection is a subset of context reasoning mechanisms that aim to select the best networks taking into account the context available. As described, context awareness procedures deal with a large amount of information, so it is critical to perform a compromise between the vital information and the helpful one. Since these schemes have to process the context data in order to provide reliable solutions and real-time decisions, a compromise may have to be made in order to select critical information so as to reduce computational effort and response time of intelligent algorithms. In this paper we focus on this kind of mechanisms so that our envisioned architecture be driven based on the context-aware selection decisions.

4 Network Selection

The network selection scheme developed aims to improve the efficiency of the multiparty content distribution based on the context information available. We decided to perform access and core network selection separately, since the involved constraints for both parts are different. Moreover, a unique algorithm involving all parameters would be infeasible and most of all not scalable. The access selection scheme is essential given the heterogeneous and multihomed aspects of FNs, attempting to connect terminals to the access networks that best suit their communication needs. To accomplish this, NUM comprises users' context, session requirements and network context in its access selection mechanism. Based on the access network selected, the core network path selection is performed to determine the best feasible path between session content source and the selected AP, considering network QoS metrics and available bandwidth.

The selection scheme comprehends both access network choice and core network path, in which the last one depends on the access network selected. This strategy attempts to use the context information available towards a more efficient and personalized multimedia content delivery system. As depicted in Figure 5, our approach considers two different schemes for Access and Core network selection, given that the intended objectives and constraints are not the same. Both procedures are executed on NUM, starting with the access network selection to obtain the most suitable Access Point (AP) for the user. This procedure attempts to achieve the best compromise between user preferences/requirements and network resources, giving a special attention to the group/sub-grouping concept. Based on the AP chosen, the core network path selection is performed to determine the best feasible path between session content source and the selected AP, considering network QoS metrics, available bandwidth and expected energy consumption (e.g. influence routing in order to avoid network nodes that consume more energy).

4.1 Access Selection

The access network selection scheme adopted is described by its pseudo-code in Algorithm 1. It was designed to determine a ranked list of candidate access points (APs), with sufficient resources and to which the terminal is able to connect. Since the architecture proposed in this work aims at the support of a context-aware multiparty content delivery, it would be on the best interest of operators to aggregate in the same AP users that share the same content source, CoS or other features (e.g. device capabilities, location, codecs).

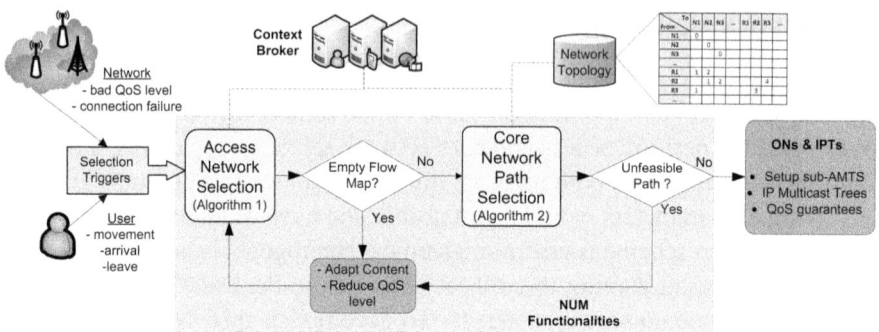

Figure 5 Overall context-aware selection scheme.

Algorithm 1: Access network selection algorithm.

$AP\_CANDS = \{AP_1, \ldots, AP_N\}$, List of candidate APs;
$Flow\_QoS \Leftarrow$ Get session QoS requirements from CB;
$M \Rightarrow$ number of APs and $W \Rightarrow$ number of constraints;
$FM = 0 \Rightarrow$ Matrix Flow Map contains APs and their rank;
for $i = 1$ to N **do**
  Get QoS context of AP_i from CB;
  if AP_i resources do not meet $Flow\_QoS$ **then**
    Remove AP_i from AP_CANDS;
  end if
end for
if $AP\_CANDS == Empty$ **then**
  Abort and re-select with different content format or CoS;
end if
Require: Matrix $APN_{M \times W}$
for $i = 1$ to M **do**
  Fill respective AP_i properties in APN;
end for
Require: Creation of Matrix $UP_{W \times W}$ according to user profile stored in CB;
$FM = APN_{M \times W} \times UP_{W \times W}$;
$FM = \sum_{i=1}^{W} FM$;
Sort FM with highest ranked AP on top;
return FM;

Here we propose to include in the access selection process the AP constraints depicted in Table 1: User Preference, Energy Consumption, Subgrouping, CoS and Bandwidth Allocation. Nevertheless, a wide variety of parameters could also be used, such as signal strength, Bit Error Rate, location or access cost. Table 1 values span between 0 and 100, the higher representing the better value. The User Preference value of each AP is empirically assigned by the user, possibly taking into account its background

Table 1 AP Properties and respective values.

Access Technology	User Preferences (static)	Energy (static)	Sub-grouping (dynamic)	CoS (dynamic)	Bandwidth Allocation (dynamic)
WiMAX 1	100	30	90	90	50
UMTS 1	70	50	60	50	100
Wi-Fi	80	80	0	0	90
WiMAX 2	100	30	30	70	10

Table 2 Weight distribution according to user profiles.

User Profile	User Preferences	Energy	Sub-grouping	CoS	Bandwidth Allocation
Business Man	0.5	1.0	1.5	1.0	1.0
Gamer	1.5	0.5	1.0	1.5	1.0
Groupie	0.5	1.5	0.5	0.5	0.5

experience. The Energy parameter is associated with the level of energy consumed when accessing the network through that access point. As stated on the related work, technologies able to provide higher data access rates consume more energy. However, a user with a low level of battery may not be willing to connect to that kind of access, preferring an access that could save more energy. Bandwidth Allocation is the result of a simple function that returns the value 100 when any traffic flows through the AP, and 0 if there are no resources available. Considering the sub-grouping constraint, a higher value means that the AP is serving a low number of different groups and it may support a session of a different group. The CoS parameter has a similar function, as it aims to aggregate traffic with the same CoS, being assigned higher values for APs that already have flows with the same CoS. Table 2 holds, for each user profile, the weights that each type of user assigns to the constraint involved in the access selection. These values span between 0 and 1.5, allowing constraint differentiation according to each user. In the parameters that compose the set of user context, energy refers to the energy constraints of the user (for example, using a mobile phone with low battery), and the sub-grouping refers to the ability to be grouped with other users for content delivery.

The access selection algorithm, Algorithm 1, starts by obtaining from the CB a list of APs (*AP_CANDS*) in the range of the user. From the same repository, it also gets the session QoS demands (*Flow_QoS*) mapped in bandwidth, delay and loss rate requirements. This information is then crossed with the QoS context of each AP candidate, executing a simple admission control mechanism. Then, the matrix *APN* is built based on the context of each AP that passes the admission mechanism, which final form will be similar to Table 1. After that, it is also created the matrix *UP* with the user context relative to each of the constraints involved (based on Table 2). Combining both matrices, and after some simple matrix operations, it is obtained the Flow Map (*FM*) with the information about the most suitable APs regarding user and communication context.

4.2 Core Selection

Once selected the most suitable AP regarding communication needs and AP and user context, the best path between content source and the chosen AP is selected. In our architecture, this path is mapped in an AMT, formed by a set of sub-AMTs bounded by ONs, where the AP is also considered as a ON. We assume that the ONs are already defined in the network, being the sub-AMTs defined by the ONs that belong to the path selected. If we consider a network where all core nodes are capable of performing ON functionalities, we may then select them and construct sub-AMTs based on that, which would require a new intelligent selection scheme that is out of the scope of this work. However, we assess the impact of ONs' number and positioning in the network performance.

Algorithm 2: Core network path selection algorithm.

$L = \{L_1, \ldots, L_N\}$, List of available links;
(L_i={Cost, BW, Delay, Weight})
Normalization Factors: NF_C, NF_B, NF_D
$dst \Rightarrow$ List of Egress ONs to which traffic is intended;
$Flow\_QoS \Leftarrow$ Get session QoS requirements from CB;
for $i = 1$ to N **do**
    if (L_i.BW<$Flow\_QoS$.BW)&&(L_i.Delay>$Flow\_QoS$.Delay) **then**
        Remove L_i from L;
    else
        L_i·Weight = L_i·Cost $\times NF_C + L_i$·BW $\times NF_B + L_i$·Delay $\times NF_D$;
    end if
end for
if $L == Empty$ **then**
    Abort and re-select with different content format or CoS;
end if
$Path = Dijkstra(Session\ Source,\ dst)$
Require: Sub-AMTS Definition $\Rightarrow$ Divide Path according to ONs;
Search for already defined common Sub-AMTs;
return List of new Sub-AMTs;

The core path selection scheme is presented in Algorithm 2. Available links in the network (L) and their characteristics in terms of bandwidth available (BW), routing cost ($Cost$) and mean delay ($Delay$) observed are retrieved from NUM's database, being essential that this information be updated before performing a selection. Another crucial input is a list (dst) of Egress ONs to which the content source will send traffic, including not only the AP previously selected but also all the other APs that share the content source. Once again, considering session QoS demands ($Flow\_QoS$) retrieved from the CB, an admission control is executed based on links' QoS context. To each admitted link, it is computed a cost ($Weight$) combining link's bandwidth available, delay and routing cost. Normalization factors (NF_C, NF_B, NF_D) are associated to each of these parameters so that links' weights may be related with each other. Dijkstra's algorithm is then employed to compute the constrained shortest path to the respective APs. Based on the resultant path, and considering ONs within it, sub-AMTS are defined between two ONs. However, if there is a path previously selected for the same content source, some sub-AMTs already defined are very likely to be re-used, which saves network and resource configurations.

In both selection schemes, a situation can occur in which the network resources available are not sufficient to accommodate a new flow with its QoS required. In order to solve this issue, a possible solution could be the re-establishment of a new session with a different content format (codec) requiring less resources. At the network level, the session content could be distributed with a lower level of QoS supported by network resources.

5 Results

The architecture described so far was implemented in the Network Simulator 2 (NS-2) [18]. In order to assess its performance and robustness, it was created a flexible network topology regarding the number of ingress, egress and core nodes, core ONs, APs, Mobile Terminals (MTs), data sources and sessions. This way, we easily evaluate the response of the implemented scheme by varying the input parameters and generating different scenarios. All experiments were based on the scenario of Figure 6, with nine network nodes, changing some specific parameters according to the evaluation test characteristics. The links in the core network were configured with a random delay between 1 and 2 ms and a bandwidth ranging between 5 and 7 Mbps. Although the nodes are fixed, the links between the nodes are randomly generated.

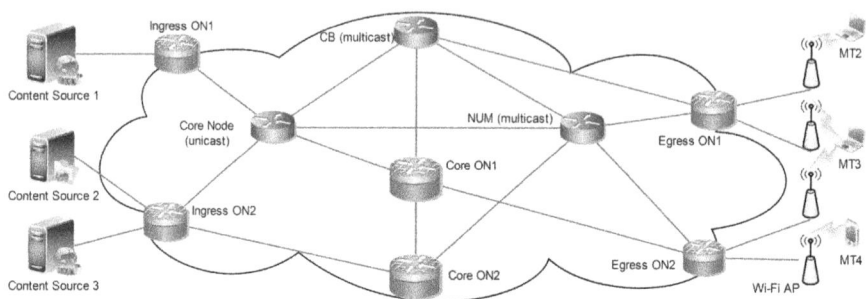

Figure 6 General evaluated scenario.

The implementation revealed several limitations of the NS-2 regarding the incompatibility between multicast (flat routing), domains (hierarchical addressing) and scenarios involving wireless and fixed parts. To overcome these issues, we decided to emulate the wireless connections onto wired links. Thus, we employ a dynamic link error model, adapting delay and losses, according to previous simulations made on wireless scenarios. Moreover, the distances between the APs were defined so that their coverage areas do not overlap and the communications do not interfere between each other. Two different traffic generators were randomly used: exponential traffic and constant bit rate. All flows have a packet size of 1000 bytes and an average rate of 100 Kbps. Regarding the QoS implemented mechanisms, 6 CoS were considered: signalling, routing, Expedited Forwarding, Best Effort (BE), Assured Forwarding 1 (AF1) and AF2. The CoS of each flow is randomly chosen, but it is guaranteed that more than a half of the traffic is BE to simulate a real network.

5.1 Access Network Selection Methods

This evaluation shows the performance response of different access network selection methods. The "Basic" method represents the traditional AP selection based on the signal strength. The second one, "Sub-grouping Without CoS", treats all traffic as equal using a load balancing mechanism, user preferences and cost. The last method, "Sub-grouping With CoS", introduces the Sub-grouping and CoS in order to better manage traffic with different requirements. It also provides an admission control mechanism to prevent high priority traffic to be impaired.

The input parameters are the ones presented in the previous tables. Considering the results obtained for the different access strategies, Figure 7 shows

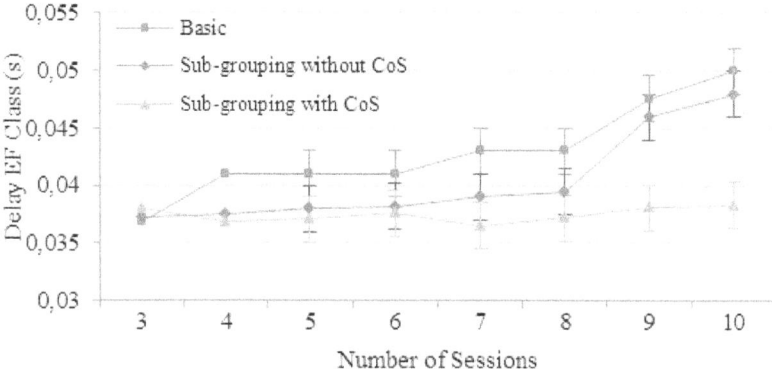

Figure 7 EF class delay for different access selection methods.

that the method "Sub-grouping with CoS" is obviously the better choice when dealing with expedited forwarding DiffServ class of traffic. Moreover, the delay value maintains nearly constant due to the above mentioned admission mechanism. Comparing the "Basic" approach with the "Sub-grouping without CoS" method, we may observe better delays for each one of the sessions' number simulated. These results show that, intelligently involving more information in the access selection process, it is possible to achieve better performances while maintaining user satisfaction.

5.2 Sub-Grouping

The sub-grouping concept already described in section III.A allows two or more users to receive the same content but with different formats in order to a better adaptation to user capabilities and network conditions. From a multicast perspective, users are in the same group but with different sub-groups. In our implementation, sub-grouping is also applied when a multicast group is divided between different APs in order to maintain the QoS correspondent to each CoS.

In this evaluation, users have different preferences for the sub-grouping parameter, being the network performance evaluated for different values of the sub-grouping weight. A high sub-grouping value means more users receiving content in the same AP within the same group.

According to Figure 8, as the sub-grouping weight increases, the network load decreases, which is more evident for a number of sessions higher than 3. Despite not presented, the overhead has a similar behavior. This is explained

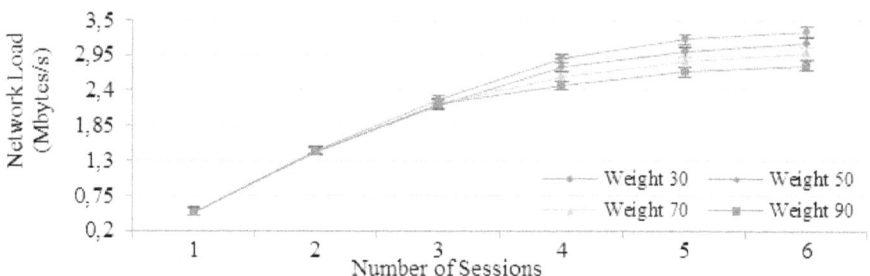

Figure 8 Network load varying sub-grouping weight.

by the fact that with a high value assigned to the sub-grouping property, the same AP has more chances to be chosen and consequently the same core path. Thus, it is avoided unnecessary content replication, as long as the QoS requirements are satisfied. In fact, the algorithm should be designed to achieve a good compromise between acceptable delays and network load.

6 Conclusions

Given the recent concerns about sustainability and environmental footprint, we presented in this work how the heterogeneity of FNs can represent a source of rich information towards the development of algorithms and mechanisms that may increase the overall communication efficiency, taking into account energy requirements and constraints. The well-known concept of context-awareness is one of the most suitable features of next generation architectures to overcome the diversity of technologies and endow networks with the ability to adapt according to changing conditions. This paper presents an architecture for personalized multiparty content delivery with a special focus on the context-aware selection mechanisms, both for access and core networks. Two different network selection schemes were presented, focused on choosing the best networks according to network, user and surrounding context. Moreover, it should be stressed the flexibility of the developed selection algorithms so that any kind of constraint could be involved in the selection process. The sub-grouping functionality implemented, despite very simple, provides good results when avoiding data redundancy while ensuring a certain level of personalization. The results obtained through the simulation allow to conclude that these mechanisms are able to not only decrease network overall delay, but also provide a more scalable solution considering necessary network re-arrangements and avoiding the necessity of global re-configurations, thus allowing green networking by reducing overall energy consumption. The work developed allows us to conclude that context driven architectures are more suitable to face FNs challenges, providing higher levels of adaptation and personalization, enabling the development of novel applications that may be an added value as for operators as for consumers. Moreover, with the proposed approach, it was provided the integration of energy as one of the inherent context parameters to be used in the network operation and control.

7 Future Work

Regarding future developments for the architecture presented in this paper, we plan to develop both selection algorithms, by introducing self-management and autonomic concepts to achieve self-organized architectures in order to achieve a scalable solution able to cope with the complexity and constant changes of future communication environments. We also aim to evolve the current control strategy towards a decentralized system, which will raise new challenges in what concerns the distribution of the intelligence and knowledge throughout the network elements. Following autonomic and cognitive principles, we aim to develop a cognitive cycle within the main elements of our architecture in order to incorporate intelligent decision mechanisms performing network optimization at different levels. Our investigation will focus on closing the management loop, mainly performing informed management decisions to optimize a pre-defined goal. The capabilities developed must be distributed throughout some network elements in order to address the complexity issues introduced by context-awareness in heterogeneous environments. Although being able to include the energy requirements and constraints in this self-management and distributed process, our aim is for this approach to also increase the energy efficiency, due to its autonomous and distributed nature.

References

[1] Farooq Bari and Victor C.M. Leung. Automated network selection in a heterogeneous wireless network environment. *IEEE Network*, 21(1):34–40, January/February 2007.
[2] A. Bianzino, C. Chaudet, D. Rossi, and J. Rougier. A survey of green networking research. *IEEE Communications Surveys & Tutorials*, 2010.

[3] Thomas Buchholz and Michael Schiffers. Quality of context: What it is and why we need it. In *Proceedings of the 10th Workshop of the OpenView University Association (OVUA03)*, 2003.
[4] J. Chabarek, J. Sommers, P. Barford, C. Estan, D. Tsiang, and S. Wright. Power awareness in network design and routing. In *Proceedings of 27th Conference on Computer Communications, IEEE (INFOCOM 2008)*, pages 457–465, 2008.
[5] Tao Chen, Honggang Zhang, Zhifeng Zhao, and Xianfu Chen. Towards green wireless access networks. In *Proceedings of 5th International ICST Conference on Communications and Networking in China (CHINACOM)*, pages 1–6, 2010.
[6] L.M. Correia, D. Zeller, O. Blume, D. Ferling, Y. Jading, I. Go anddor, G. Auer, and L. Van Der Perre. Challenges and enabling technologies for energy aware mobile radio networks. *IEEE Communications Magazine*, 48(11):66 –72, 2010.
[7] Anind K. Dey and Gregory D. Abowd. Towards a better understanding of context and context-awareness. In *HUC 99: Proceedings of the 1st International Symposium on Handheld and Ubiquitous Computing*, pages 304–307. Springer-Verlag, 1999.
[8] EU Funded Research Project FP7. Earth (energy aware radio and network technologies). http://www.ict-earth.eu/, 2010.
[9] GreenTouch Project. http://www.greentouch.org/.
[10] Tao Gu, Xiao Hang Wang, Hung Keng Pung, and Da Qing Zhang. An ontology-based context model in intelligent environments. In *Proceedings of Communication Networks and Distributed Systems Modeling and Simulation Conference*, pages 270–275, 2004.
[11] E. Gustafsson and A. Jonsson. Always best connected. *IEEE Wireless Communications*, 10(1):49–55, February 2003.
[12] Suk Yu Hui and Kai Hau Yeung. Challenges in the migration to 4G mobile systems. *IEEE Communications Magazine*, 41(12):54–59, December 2003.
[13] A. Iera et al. An access network selection algorithm dynamically adapted to user needs and preferences. In *Proceedings of 17th Annual IEEE International Symposium on Personal, Indoor and Mobile Radio Communications (PIMRC'06)*, Helsinki, Finland, September 2006.
[14] R. Jain. Quality of experience. *IEEE Multimedia*, 11(1):96–95, 2004.
[15] V. Jesus, S. Sargento, and R.L. Aguiar. Any-constraint personalized network selection. In *Proceedings of 19th Annual IEEE International Symposium on Personal, Indoor and Mobile Radio Communications (PIMRC'08)*, Cannes, France, September 2008.
[16] B. Mathieu et al. Self-management of context-aware overlay ambient networks. In *IFIP/IEEE IM*, pages 749–752, 21 2007-Yearly 25 2007.
[17] A. Neto, E. Cerqueira, A. Rissato, E. Monteiro, and P. Mendes. A resource reservation protocol supporting qos-aware multicast trees for next generation networks. In *Proceedings of IEEE Symposium on Computers and Communications (ISCC'07)*, pages 707–714, July 2007.
[18] The Network Simulator NS-2. http://www.isi.edu/nsnam/ns/.
[19] C. Prehofer, N. Nafisi, and Q. Wei. A framework for context-aware handover decisions. In *Proceedings of 14th IEEE International Symposium on Personal, Indoor and Mobile Radio Communications (PIMRC2003)*, Volume 3, Beijing, China, September 2003.
[20] Fernando M.V. Ramos, Richard J. Gibbens, Fei Song, Pablo Rodriguez, Jon Crowcroft, and Ian H. White. Reducing energy consumption in iptv networks by selective pre-

joining of channels. In *Proceedings of the First ACM SIGCOMM workshop on Green Networking*, pages 47–52, 2010.
[21] Oriana Riva et al. A next generation operator environment to turn context-aware services into a commercial reality. In *Proceedings of International Conference on Mobile Data Management*, 2008.
[22] B.N. Schilit and M.M. Theimer. Disseminating active map information to mobile hosts. *IEEE Network*, 8(5):22–32, September/October 1994.
[23] Thomas Strang and Claudia Linnhoff-Popien. A context modeling survey. In *Proceedings of Workshop on Advanced Context Modelling, Reasoning and Management, UbiComp 2004 – The Sixth International Conference on Ubiquitous Computing*, Nottingham, England, 2004.
[24] M.H. Williams et al. Context-awareness and personalisation in the daidalos pervasive environment. In *ICPS*, pages 98–107, July 2005.
[25] B. Xing and Nalini Venkatasubramanian. Multi-constraint dynamic access selection in always best connected networks. In *Proceedings of the Second Annual International Conference on Mobile and Ubiquitous Systems: Networking and Services (MobiQuitous2005)*, pages 56–64, July 2005.

Biographies

Nuno Coutinho concluded his five-years Integrated MSc in Electronics and Telecommunications Engineering at University of Aveiro in 2008. His master's dissertation, entitled "Intelligence in Mobility Decisions", was about the selection of the best access network according to context information about user and network. Since October 2008 he is a PhD student at University of Aveiro and he joined the Celfinet Innovation Department from October 2008 to February 2009. Currently, he is a researcher associated to the Institute of Telecommunications, being involved in National projects (MuMoMgt, GEN-CAN, UbiquiMesh) and European projects (C-CAST). His main research interests are related to Heterogeneous Networks, Multicast, Overlay Networks, context-awareness and Quality-of-Experience.

Susana Sargento (http://www.av.it.pt/ssargento) received her PhD in 2003 in Electrical Engineering. She joined the Department of Computer Science of the University of Porto in September 2002, and is in the University of Aveiro and the Institute of Telecommunications since February 2004. She is also a Guest Faculty of the Department of Electrical and Computer Engineering from Carnegie Mellon University, USA, since August 2008, where she is currently performing Faculty Exchange. She has been involved in several national and European projects, taking leaderships of several activities in the projects, such as the QoS and ad-hoc networks integration activity in the FP6 IST-Daidalos Project. She has been recently involved in several FP7 projects (4WARD, Euro-NF, C-Cast, WIP, Daidalos, C-Mobile), national projects, and CMU—Portugal projects (DRIVE-IN with the Carnegie Melon University). She has been TPC-Chair and organized several conferences, such as MONAMI'11, NGI'09, IEEE ISCC'07. She has been also in the Technical Program Committee of several international conferences and workshops (more than 20 in 2009), such as ACM MobiCom 2009 Workshop CHANTS and IEEE Globecom (2010). Furthermore, she has been a reviewer of numerous international conferences and journals, such as *IEEE Wireless Communications*, *IEEE Networks*, *IEEE Communications*, *Telecommunications Systems Journal*, *IEEE Globecom*, *IEEE ISCC* and *IEEE VTC*. Her main research interests are in the areas of Next Generation and Future Networks, more specifically QoS, mobility, self- and cognitive networks. She regularly acts as an Expert for European Research Programmes.

Tiago Condeixa (http://www.av.it.pt/tcondeixa) concluded his Integrated Msc in Electronic and Telecommunications Engineering at University of Aveiro in June of 2009. His Msc Thesis, entitled "Evaluation of the Control of Multicast Sessions in Context-Aware Networks", was integrated in the FP7 project C-Cast. He was awarded with a merit scholarship for being one of the best students of University of Aveiro in 2008/2009. Currently, he is a PhD Student of University of Aveiro, Portugal, since October of 2009. He is also a researcher in the Institute of Telecommunications of Aveiro, being involved in a National project (User-centric Mobility Management) and European project (C-Cast). His main research interests are related with Future Internet, more precisely Mobility Management, User-centricity, Context-awareness, QoE, Intelligent Networks and Multicast. He is also interested in Network Simulator 2 and 3.

Augusto José Venâncio Neto is an Associate Professor at the Teleinformatics Engineering Department of Federal University of Ceara. Moreover, he is member of Group of Computer Networks, Software Engineering and Systems (GREat) at UFC, Brazil, as well as of the Telecommunications Institute at University of Aveiro, Portugal. He got his B.Sc. on Technologist in Data Processing from the University of Amazonia (1996), and his MSc in Computer Science from Federal University of Santa Catarina (2001). He got his PhD from the University of Coimbra (UC), Portugal, in 2008, working under the SAPRA and Q3M European projects. He made his first post-doc at the Institute of Telecommunications (IT), Polo University of Aveiro (Portugal), funded by the European and Portuguese projects. He regularly serves as reviewer for several conferences and journals in the field of networking and telecommunications. He was a visiting professor at the Polytechnic Institute of Coimbra. His current research interests are on Next Generation Networks, Future Internet, Quality of Service (QoS), Quality-Oriented Routing, Multicast (IP Multicast and Overlay Application Layer Multicast), Broadband mobile access technologies and Smart Grid.

Online Manuscript Submission

The link for submission is: www.riverpublishers.com/journal

Authors and reviewers can easily set up an account and log in to submit or review papers.

Submission formats for manuscripts: LaTeX, Word, WordPerfect, RTF, TXT.
Submission formats for figures: EPS, TIFF, GIF, JPEG, PPT and Postscript.

LaTeX

For submission in LaTeX, River Publishers has developed a River stylefile, which can be downloaded from http://riverpublishers.com/river_publishers/authors.php

Guidelines for Manuscripts

Please use the Authors' Guidelines for the preparation of manuscripts, which can be downloaded from http://riverpublishers.com/river_publishers/authors.php

In case of difficulties while submitting or other inquiries, please get in touch with us by clicking CONTACT on the journal's site or sending an e-mail to: info@riverpublishers.com

www.ingramcontent.com/pod-product-compliance
Ingram Content Group UK Ltd.
Pitfield, Milton Keynes, MK11 3LW, UK
UKHW021321180426
11947UKWH00015B/1359